The Culture of Feedback

The Culture of Feedback

Ecological Thinking in Seventies America

DANIEL BELGRAD

The University of Chicago Press   Chicago and London

The University of Chicago Press, Chicago 60637
The University of Chicago Press, Ltd., London

For more information, contact the University of Chicago Press, 1427 E. 60th St., Chicago, IL 60637.
Published 2019
Printed in the United States of America

28 27 26 25 24 23 22 21 20 19 1 2 3 4 5

ISBN-13: 978-0-226-65236-8 (cloth)
ISBN-13: 978-0-226-65253-5 (paper)
ISBN-13: 978-0-226-65267-2 (e-book)
DOI: https://doi.org/10.7208/chicago/9780226652672.001.0001

Library of Congress Cataloging-in-Publication Data

Names: Belgrad, Daniel, author.
Title: The culture of feedback: ecological thinking in seventies America / Daniel Belgrad.
Description: Chicago: The University of Chicago Press, 2019. | Includes bibliographical references and index.
Identifiers: LCCN 2019000821 | ISBN 9780226652368 (cloth: alk. paper) | ISBN 9780226652535 (pbk: alk. paper) | ISBN 9780226652672 (e-book)
Subjects: LCSH: Ecology—United States—Philosophy—History—20th century. | Environmentalism—United States—History—20th century.
Classification: LCC QH540.5 .B45 2019 | DDC 577.0973—dc23
LC record available at https://lccn.loc.gov/2019000821

♾ This paper meets the requirements of ANSI/NISO Z39.48-1992 (Permanence of Paper).

Contents

Acknowledgments

In conversations with my friends Mike Witmore, Ying Zhu, and Dell DeChant, I first encountered words and ideas that grew in importance as I came to write this book. At annual conferences of the Cultural Studies Association and as a visiting lecturer at the Philosophy Department of the University of Central Florida, I was given opportunities to benefit from the input of responsive audiences to early versions of my argument. Sandra Law and Ginny Gates-Fowler at the University of South Florida Library kept me supplied with a constant stream of crucial but hard-to-find sources, and Richard Schmidt helped me to gather the illustrations that I needed. I am especially grateful for the community of my colleagues in the Humanities and Cultural Studies Department at the University of South Florida, where Brendan Cook and Amy Rust read the entire book in manuscript and gave me their constructive criticisms. Outside the department, Fred Turner and John Wilson did me the same favor. Doug Mitchell, my editor at the University of Chicago Press, faithfully waited twenty years for this manuscript. Kyle Wagner and Erin DeWitt helped me to finalize it. I owe so much to my daughters, Lydia Lutsyshyna and Liz Valentine, and most of all to my wife, Catherine Valentine. They carry me forward.

Introduction

We're just a biological speculation
Sitting here, vibrating
And we don't know what we're vibrating about . . .
Oh, and if and when the law of man
Is not just, equal and fair
Then the laws of nature will come and do her thing.
FUNKADELIC, "BIOLOGICAL SPECULATION"
FROM *AMERICA EATS ITS YOUNG* (1972)

We speak casually of improving a course of action by getting some feedback, as if that were the most natural thing in the world. But the idea of feedback itself has a history. During the Second World War, "feedback" developed as a term to refer to the dynamics of self-regulating mechanical systems, which correct their actions by "feeding" some effects "back" into the system as input to influence later actions. Due to the ability of such systems to self-correct, or "learn," they could be considered intelligent.

Conversely, systems theory, which developed to describe how such systems worked, came to define intelligence itself as the ability to self-correct in response to feedback. Redefining intelligence this way—not as a uniquely human faculty produced by consciousness, but as the property of a system governed by feedback loops—eventuated in new ways of thinking about the varieties of intelligence found in nature. This is what I mean by ecological thinking.

Variously termed "ecology" or "cybernetics," the vision of the intelligent system governed by feedback loops has been a key image in American culture since the middle of the twentieth century. The term "cybernetics" is most often associated with the creation of artificial feedback systems using electrical circuits (that is, artificial intelligence). Foregrounding the term "ecology" instead connotes a focus on natural systems. The "new ecology" that developed in the 1960s described the complex feedback loops that define interactions among plants, animals, and their physical environments (fig. 1).[1]

The idea of feedback dynamics as the connective tissue of systems, broadly conceived, became widespread in American intellectual discourse in the postwar decades, finding expression in academic specializations as diverse as engineering, sociology, biology, and psychology. Yet it was not until the early 1970s that a distinct popular culture emerged around the concept, as feedback became the governing trope for a counterculture that reoriented and reinvigorated the postwar culture of spontaneity and the psychedelic sixties. The idea that systems-based forms of intelligence were ubiquitous in the natural environment captured the public imagination during the seventies, resulting in the creation of new and widespread cultural practices. In this sense, ecological thinking is not identical with the science of ecology. As David Oates wrote in his study *Earth Rising: Ecological Belief in an Age of Science,* "ecological thinking" refers to "the pattern of thinking called ecology [that] has proved to have potentials far beyond the science itself . . . [because] the ecological worldview is not a science: it is a belief system extrapolated from one."[2]

Ecological thinking includes environmentalism but is not limited to it. Environmentalism emphasizes the fact that we humans are nested within nature's complex systems. Therefore we must interact with those systems in ways that do not jeopardize our own survival. But ecological thinking extends to a social vision as well. Thus in 1973 philosopher Arne Næss summarized the principles of what he called "deep ecology" as opposed to a "shallow," merely environmentalist focus, writing: "There are deeper concerns which touch upon principles of diversity, complexity, autonomy, decentralization, symbiosis, egalitarianism, and classlessness."[3]

Many Americans articulated the connections between their ecological thinking and the practices they derived from them, while many more engaged in those practices without overtly parsing their logic. This is in the nature of cultural forms as the vehicles of ideas, because practices convey meanings by embodying them rather than announcing them, using them to organize and shape one's actions. Central to the construction and communication of meaning through patterns

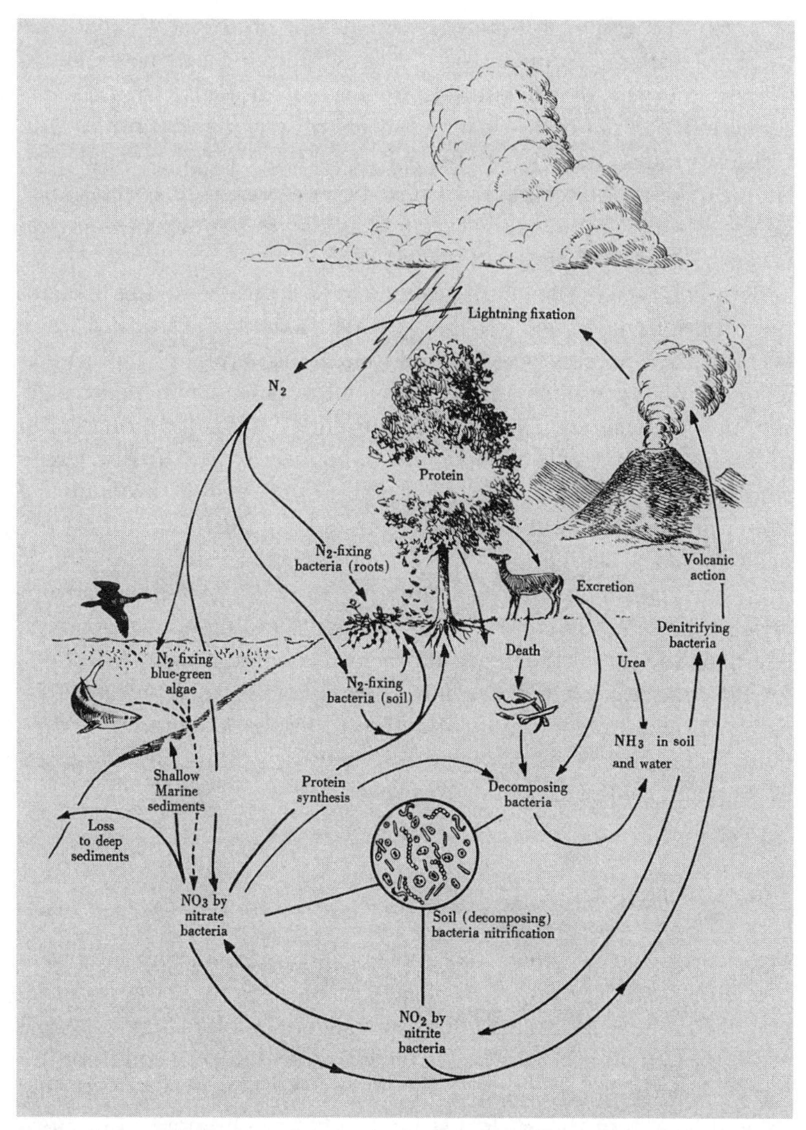

Figure 1. An example of an ecosystem showing feedback loops (the nitrogen cycle), originally from Robert L. Smith's *The Ecology of Man: An Ecosystem Approach* (1972). (Robert L. Smith, *Ecology and Field Biology*, 5th ed., © 1996. Reprinted by permission of Pearson Education, Inc., New York, New York.)

of practice is the evocation and articulation of affective experience through aesthetic form. By such means, thinking can be extended beyond the pale of recognizable intellectual discourse. In this way sometimes popular culture can constitute the cutting edge of thought. Conservationist Paul Shepard wrote in 1969 that the "greater and overriding wisdom [of ecology] . . . can be approached mathematically, chemically, or it can be danced or told as a myth." For epistemological reasons, he actually preferred the last two.[4]

This book traces the manifestations of ecological thinking in a culture of feedback throughout the seventies. Its scope extends to wherever feedback systems were invoked to imagine natural phenomena as forms of intelligence, or "mind." The result is not a comprehensive description of American culture in the seventies, nor is it a full account of the use of the feedback metaphor in intellectual and artistic circles since its inception. Instead, it is a delineation of how an enduring American counterculture responded to and shaped the historical imperatives of the decade.

In the seventies, ecological thinking took on widespread significance because it offered a new way of understanding how to go about changing society for the better. It provided convincing new definitions for old ideals, especially those of freedom, progress, and efficiency. The ability to evolve creatively in response to feedback was embraced by many Americans as the most meaningful way to define these most important and contested political keywords.[5]

The Historiographical Context

My recovery of the influence of ecological thinking during the seventies reinforces a growing historiographical sense that it was a crucial decade in Americans' transition to postmodernity. More importantly, it gives us a better understanding of what that postmodern condition is.[6]

For about a quarter of a century after the seventies ended, published histories of the era focused on providing chronologically organized surveys of its political events and social trends. Peter Carroll's *It Seemed Like Nothing Happened: America in the 1970s* (1982) was probably the first of these; and Bruce Schulman's *The Seventies: The Great Shift in American Culture, Society, and Politics* (2001) is arguably the most influential. In 2006 there began to be published a second wave of histories whose authors absolved themselves of the earlier felt imperative to offer a comprehensive historical account of the decade. They instead organized their studies around specific themes, exploring how each one's various ramifications elucidate the texture of life in those times. Andreas Killen's *1973 Nervous Breakdown: Watergate, Warhol, and the Birth of Post-Sixties America* (2006) focuses in this manner on crises of identity. Philip Jenkins's *Decade of Nightmares: The End of the Sixties and the Making of Eighties America* (2006) describes how the discourse surrounding child abuse grew to have a powerful cultural and political presence. Thomas Hine's *The Great Funk: Falling Apart and Coming Together (on a Shag Rug) in the Seventies* (2007) emphasizes material objects, combing them for insights into the behaviors and belief systems of which they are the artifacts. My book uses the impact of ecological thinking to look at the decade from yet another angle.

The aspects of postmodernity that emerge as most salient from this perspective concern an ontological and epistemological shift away from objectivity and toward intersubjectivity and mind-body holism. The culture of feedback extended the explorations of a previous generation into what it meant to be human in the context of this alternative metaphysics.[7] Like theirs, its search for a different humanism was motivated by a sense of the dominant culture's failures and shortcomings. The embrace of objectivity formed the basis for an advanced technological mastery of nature; the culture of feedback sought to interact with nature on terms of dialogue rather than mastery. A dichotomization of mind and body was necessary in order to sustain a faith in objectivity; the culture of feedback instead espoused a definition of rationality encompassing the bodily affects and emotions. In defiance of the scientific method, the culture of feedback insisted that some necessary knowledge could only be had by learning through emulation and empathy.

As the subtitles of both Killen's and Jenkins's histories suggest, the question of what the cultural upheavals of the sixties left as their legacy haunts narratives of the seventies, as does the question of where the culture of the eighties derived from. This book sheds some light on both of these issues, as a result of its particular thematic focus. Although the bulk of the events that I write about transpired between 1971 and 1979, cultural eras seldom coincide with calendar dates. In my research, I resisted assigning arbitrary cutoff dates to the phenomena I investigated, instead letting them manifest their own boundary conditions. Repeatedly, 1983 emerged as an end date, and 1963 as the year in which an idea or practice first emerged. In using ecological thinking to tell the story of the seventies and its relationship to the years before and after, I place myself most directly in dialogue with

Fred Turner's book *From Counterculture to Cyberculture: Stewart Brand, the Whole Earth Network, and the Rise of Digital Utopianism* (2006). Turner's work focuses on the career of Stewart Brand (the publisher of the *Whole Earth Catalog*) to tell the story of cybernetics as a prehistory of the culture of neoliberalism. In the sixties, Brand embraced the possibilities of cybernetics "as an intellectual framework and as a social practice"; the *Whole Earth Catalog,* conceived as an intellectual forum rooted in systems theory for the communalist lifestyle, sold nearly a million copies between 1968 and 1972, and won (in '72) the National Book Award for Contemporary Affairs. Turner's thesis is that the attitude toward technology manifested in the *Whole Earth Catalog* reveals the masculinist, white-privileged, technocratic-consumerist, libertarian bias of the sixties counterculture. This bias, he argued, was reinforced by the cyberculture of Silicon Valley, to reemerge in the Republican political agendas of the late 1980s and 1990s. In Turner's history, Brand plays a key role, as the facilitator of an interface between the countercultural and technological communities where many early programmers situated themselves.[8]

Because I focus on ecological thinking rather than on digital cybernetics, the scope of my research has differed considerably from Turner's. From this perspective, Stewart Brand appears as a figure of only peripheral importance, as compared to such people as Gregory Bateson, Gary Snyder, Pauline Oliveros, and John Lilly. And even Brand emerges, in the discursive context I recover for him, as a more complex figure than his career as a digital utopian might suggest.

To paraphrase the poet Gary Snyder, from an ecological standpoint the fatal flaw of cyberculture is that it offers networking without community. Community grounds group dynamics in the physical and social realities of place, forcing individuals to engage in cooperation and conflict to forge a shared habitat.[9] By contrast, networking without community implies only a virtual connectedness that easily devolves into egoism. Brand himself was disillusioned with this aspect of the *Whole Earth Catalog,* as he wrote in a *Harper's* magazine article of 1973 that paid homage to the ecological ideas of Bateson. "Cybernetics is the science of communication and control. It has little to do with machines unless you want to pursue that special case. It has mostly to do with life, with maintaining circuit," Brand wrote.

> I came into cybernetics from a pre-occupation with biology, world saving, and mysticism. What I found missing was any clear conceptual bonding of cybernetic

whole-systems thinking with religious whole-systems thinking . . . [to] evoke a shareable self-enhancing ethic of what is sacred, what is right for life.

Even "three years of scanning innumerable books for *The Whole Earth Catalog* didn't turn it up." But finally, after he had given up on the *Whole Earth Catalog*, he found what he was searching for in Bateson's *Steps to an Ecology of Mind*.[10]

Brand invokes in his essay what might at first seem to be an idiosyncratic mix of influences: systems theory, biology, world saving, and religious mysticism. But I have found it to be representative of a cultural current that was, in the seventies, bigger than digital utopianism. Ten years after Brand published his *Harper's* magazine essay, one could encounter a very similar mix of references in Peter Russell's *The Global Brain* (1983), which was subtitled *Speculations on the Evolutionary Leap to Planetary Consciousness*. Russell's book weaves together systems theory, coevolution, environmentalism, synchronicity, and telepathy. It calls on its readers to prepare mentally for the leap to planetary consciousness via meditation, biofeedback, psychedelic drugs, and digital communications networks. These tools, Russell wrote, would generate "positive feedback" for people on their individual journeys to higher consciousness, which must eventually coalesce to create a "social superorganism" with one mind but diverse ecological niches.[11] Clearly, there were other destinations for systems thinking in the counterculture besides Silicon Valley.

This book traces some of those other pathways. I am interested in how ecological thinking found expression through a range of cultural practices in a variety of media. Some of the forms that it took found names: environmentalism, biofeedback therapy, ambient music, contact improvisation, and horse whispering, to cite a few. Others—such as the practice of playing music to plants—never did. But all were united by the idea that animals, plants, and even entire ecosystems embody forms of mind that we will sooner or later come to recognize as similar to our own.

My book, in tracing the development of this idea, joins a wave of recent scholarship evincing a renewed interest in it. These works include Eduardo Kohn's and Michael Marder's philosophical meditations on plant-thinking; works of posthumanism centered on critical animal studies rather than on digital media and robotics; and applications of affect theory that place the embodied and affective subject at the center of cultural meaning-making. This contemporary intellectual trend carries along with it the legacy of seventies ecological thinking, in ascribing intelligence to any "self-organizing emergent phenomena [that are] . . . 'morphodynamic'—that is, characterized by dynamics that generate form."[12]

Why It Matters: Two Ideas of Efficiency

Once it is recognized as constituting a coherent worldview, the seventies culture of feedback can be seen to take part in a long political struggle in American society over the nature of democracy and the meaning of freedom. The importance of recovering the history of this counterculture lies in its relevance to that longer history. One way of grasping the terms of the debate is by exploring the conflicting meanings attached by two discursive traditions to an important twentieth-century American keyword: efficiency. In the first decades of the twentieth century, efficiency emerged as a moral value to serve as a guide for organizing American society in the new industrial era. Increased efficiency, defined as a higher productivity ratio (more output per input), promised to produce a harmony of interests among all parts of industrial society, by simultaneously raising wages, increasing profits, and improving services. This vision of efficiency ushered in the social order known as consumer capitalism or corporate liberalism, in which incipient class conflict was attenuated through an ostensibly perpetual rise in the American standard of living.[13]

In pursuit of this vision, the scientific management techniques of Taylorism (named for industrial engineer Frederick Winslow Taylor) delegated social power to efficiency experts who were charged with finding the "one best way" of accomplishing any task by scientifically eliminating all superfluous steps in the process. Taylorism in this way helped to define efficiency as the elimination of the superfluous. Uniformly implementing the one best way entailed centralizing the decision making. Autonomy was taken away from workers on the shop floor, who were required to follow a set of precise instructions issued by engineers. Those who couldn't meet the standard would have to find some other line of work.[14]

The credo of "the elimination of the superfluous" that Taylorism embraced resonated with contemporary social Darwinist beliefs equating the progress of human society with the process of natural selection. In the social Darwinist view, extinction was seen to improve the efficiency of nature by eliminating species and bloodlines that had competed unsuccessfully for survival. Social relations were believed to mimic the natural order by awarding survival to the victors. In social Darwinist terms, for instance, the history of the American frontier was cast as the "winning of the West": a national myth confirming the superiority of European Americans to the Native Americans whom they had successfully displaced.[15]

During the crisis of consumer capitalism produced by the Great Depression and Dust Bowl of the 1930s, however, there emerged a competing vision of efficiency. This associated it with the optimal circulation of all resources available to a particular community. In contrast to the Taylorist principles of maximum productivity and the elimination of the superfluous, the ecological idea of efficiency emphasized conservation, inclusivity, and a close attention to the dynamics of community interaction.

The technique of contour plowing promoted by New Deal agronomists Paul Sears and Aldo Leopold illustrates the ecological principle (fig. 2). In the Taylorist view, the most efficient furrow was the straightest one, comprising the shortest distance between two points and the

Figure 2. Contour plowing. Photograph by Joe Munroe. Courtesy of the Ohio History Connection.

simplest method for covering a rectangular plot of land. According to Sears and Leopold, however, the extra time and effort required by the farmer to keep the furrow constantly perpendicular to the slope of the land was an investment in community that ultimately paid off in improved soil and water conservation and an increased crop yield. On this basis, Sears counseled farmers in 1935, "Do not . . . indulge in the *vanity* of straight furrows. Plow with the contour of the land." Contour plowing takes into account how the diverse components of the ecological community—the topography, the topsoil, the plants, the rainfall, and the farmer—respond to their interactions, and it associates efficiency with this holistic viewpoint.[16]

Leopold and Sears saw this agronomic principle as just one example of the general lesson that human societies must be built on a respect for the dynamics of community interaction. Leopold observed in 1933, "Civilization is not as [the historians of progress] often assume, the enslavement of a stable and constant earth. It is a state of mutual and interdependent cooperation between human animals, other animals, plants, and soils." An ecological mind-set, Sears wrote in 1935, is intent on optimizing *"not merely what is there, but what is happening there."* It fosters mutual interdependencies. In order to check the accelerating soil erosion of the Dust Bowl, for instance, humans would have to learn to rely on "the delicate, thread-like roots of plants."[17]

The ecological idea of efficiency contradicted the conclusions of social Darwinism and Taylorism concerning the nature of progress and the value of diversity. In ecological systems, higher rates of efficiency are associated with greater complexity, not greater simplicity. Rather than standardization (the "one best way") and the elimination of the superfluous, ecological thinking values diversity and inclusivity. Diversity enables more forms of interdependency and greater resilience. Progress proceeds not by eliminating the "unfit," but by maximizing the flexibility to respond to variable conditions. Thus Sears argued in 1935 that the diversity of plant life on the arid Great Plains had provided a buffer against the region's potentially extreme climatic conditions. Plowing the plains had destroyed that buffer and "released the forces of wind and water which had been held in check . . . by a continuous carpet of plant life."[18]

These two definitions of efficiency also imply different conceptions of the individual self. Imagining life as a competition in which the "superfluous" are progressively eliminated focuses attention on the self as a discrete unit, emphasizing the boundaries that separate one person from another. It is consistent with the ideological construction of self that Anthony Wilden in *System and Structure* called "the Lockean ego." This is a self that is "autonomous in its essence" and the prototypical and most personal form of "private property."[19] By contrast, ecology's emphasis on mutualism foregrounds the intersubjective quality of selfhood. The self is reconstituted continually through one's interactions with others. Paul Shepard wrote in 1969:

> We are hidden from ourselves by habits of perception. . . . Our language, for example, encourages us to see ourselves—or a plant or animal—as an isolated sack, a thing, a contained self. Ecological thinking, on the other hand, requires a kind of vision across boundaries. The epidermis of the skin is ecologically like a pond surface or a forest soil, not a shell so much as a delicate interpenetration.[20]

Arne Næss asserted similarly in 1973 that deep ecology implied a "*relational, total-field image.* Organisms as knots in the biospherical net or field of intrinsic relations."[21] In ecological thinking, the boundaries separating the self from the system that sustains it are permeable. To some extent their designation is arbitrary.[22]

If the New Deal is seen as a political program specific to the Great Depression, it is easy also to see it as a done deal—a set of policies and accomplishments now receding into the distant past. But it is more accurately viewed as one moment in a longer continuum of American attempts to articulate an alternative social vision to the dominant liberal order, by emphasizing interdependence and redefining how progress is measured.[23] In that sense the New Deal is still ongoing. The seventies culture of feedback continued further down that path.

The two definitions of efficiency that derive from the worldviews of Taylorism and ecological thinking imply different outlooks regarding the benefits of centralized authority structures. While the self, in ecological thinking, is understood as a "subsystem" rather than a separate entity, it is, notwithstanding, a "knot" never fully subsumed into its surroundings. As Paul Weiss wrote in *The Science of Life: The Living System—A System for Living,* published in 1973, the survival of a system hinges on its subsystems' having the necessary degree of freedom to adapt creatively to their changing environments.[24] Ecological thinking therefore places a premium on organizational dynamics that preserve relative autonomy and maximize feedback; centralized authority structures that fail to do so weaken the social order. Taylorism, by contrast, endorses the centralization of power on the basis that it strengthens the social order by achieving a more efficient coordination of disparate individual energies in pursuit of the collective good. Taylorism and consumer capitalism embraced centralized power as the means to provide a higher standard of living for the average American. What could be more democratic than that? During the Second World War, American thinkers grappled with this very question.

The Historical Context

The culture of feedback's engagement with the ideas of freedom, democracy, and decentralization harks back to the origins of systems theory in the context of World War II. America's mobilization for the Second World War intensified the push for economic and political centralization in service of the war effort. At the same time, however, the centralization of power in the Soviet and Nazi dictatorships—manifested in social, economic, and even cultural directives emanating from their central authorities—stood as a potent symbol of the wrongness of their politics. Anthropologists Margaret Mead and Gregory Bateson spoke out vociferously against applying such social engineering techniques in this country even in wartime, insisting that they were antithetical to democratic values. Bateson argued in 1942 that "a basic and fundamental discrepancy exists between 'social engineering,' manipulating people in order to achieve a planned blue-print society, and the ideals of democracy. . . . It is hardly an exaggeration to say that this war is ideologically about just this."[25]

Bateson and Mead insisted that democracy had to be embraced as the *means* as well as the ends of the American war effort. Bateson asserted, "If we go on defining ends as separate from means *and* apply the social sciences as crudely instrumental means, using the recipes of science to manipulate people, we shall arrive at a totalitarian rather than a democratic system of life." The paradoxical solution, he suggested, was that a democratic leadership must "discard purpose in order to achieve our purpose."[26] In countering Nazi and Soviet propaganda, instead of promulgating its own propaganda, the American government should concentrate on fostering democratic decision-making processes. The paradox of discarding purpose in order to achieve one's purpose in a manner not predetermined would become a central tenet of the culture of feedback.

On a practical level, Bateson's recommendation to the Council for Democracy's Committee on Public Morale involved focusing on second-order purposefulness, or, as he called it, "deutero-learning." Deutero-learning is the pattern of conduct that emerges due to what is learned about how to go about learning. For instance, if a person were to lecture before an audience and pronounce that "democracy is good; fascism is bad," the message on the level of first-order learning would be pro-democratic. The message on the level of second-order learning, however, would be the opposite, because the epistemological dynamic (what the audience learns about how to learn) is that of an authority figure telling others what to think. For democracy to be operative at the level of second-order learning, the "audience" members would have to become active participants in the dialogue and engage in a process of examining the relative merits of democracy and its alternatives, arriving at their own conclusions. Then if, on the next day, another lecturer (or the same one) were to return and announce that "there has been an error: it is fascism that is good; democracy is bad," the people's habits of deutero-learning would resist that message. The patterns of deutero-learning in a culture, Bateson believed, differed according to the priorities of every social order. The cultural work of democracy could not be achieved by the propagandistic dissemination of information, but only by encouraging deutero-learning styles cultivating open habits of mind.[27]

Bateson's ideas about culture were key to his thinking on this point. He understood cultures to be systems that related lived experiences to ideas about the world via feedback loops of learning and deutero-learning. This vision linked his anthropological work to his subsequent interest in cybernetics.

It was mathematician Norbert Wiener who introduced the word "feedback" from control engineering into the general language. In 1942, the same year that Bateson articulated his theory of deutero-learning, Wiener worked out an algorithm to enable an anti-aircraft gun to predict the future location of its target based on the pilot's previous evasive maneuvers, creating an information feedback loop that made the gun self-guiding, or "intelligent." Wiener then theorized that the feedback loop was the essence of all intelligent systems, whether biological or artificial. In 1948 he gave the emerging field a lasting name when he published the best seller *Cybernetics: Or Control and Communication in the Animal and the Machine.*[28] The application of cybernetics to the life sciences and social sciences developed from a series of conferences sponsored by the Macy Foundation beginning in May 1942 and ending in 1953, on the topic of "Feedback Mechanisms and Circular Causal Systems in Biological and Social Systems." Bateson and Mead were among those who participated. For many of the thinkers involved, as Fred Turner wrote in *The Democratic Surround*, the vision of a "cybernetic" self that "advances by participating in feedback loops" presented itself as an alternative to "totalitarian psyches and societies."[29]

Ecological thinking merged Wiener's idea of intelligent systems with the insights of population ecologists, who in the 1930s had begun to explore the dynamics of interdependency governing the sizes of animal and plant populations in a particular geographical area, or "community." Working independently from both Bateson and Wiener but during the same time period (1942–44), Aldo Leopold penned a critique of purposeful action that mirrored Bateson's condemnation of social engineering. Titled "Thinking Like a Mountain," Leopold's essay described how, in nature, intentional human interventions often led to unintended negative consequences because of a lack of attention to the dynamics of community interaction. Specifically, he criticized the long-standing practice of eliminating wolves and other predators from the open range in order to protect deer, cattle, and sheep. Although the short-term result was to have more of those animals, the long-term result was overgrazing and mass die-offs. "The cowman who cleans his range of wolves does not realize that he is taking over the wolf's job of trimming the herd to fit the range. He has not learned to think like a mountain," Leopold observed. "Hence we have dustbowls, and rivers washing the future into the sea."[30]

In his essay, Leopold identified the feedback loop (although he did not use that term) as the fundamental means of governance in ecological systems. Feedback dynamics placed constraints on the behavior of every actor in the system. Intervening in those dynamics could destabilize the entire community. As his title suggests, he imagined this network of feedback loops as a decentralized form of intelligence: the "thinking" of the mountain. He ascribed both thoughts and feelings to this system. "I now suspect that just as a deer herd lives in mortal fear of its wolves," he wrote, "so does a mountain live in mortal fear of its deer. And perhaps with better cause, for while a buck pulled down by wolves can be replaced in two or three years, a range pulled down by too many deer may fail of replacement in as many decades."[31]

Leopold died in 1948; but Bateson lived until 1980 and devoted his remaining years to exploring the implications of defining intelligence as a phenomenon of complex natural systems produced by networks of feedback loops. Around 1967 he adopted ecology as a central metaphor. His work of three decades was collected in the book of essays *Steps to an Ecology of Mind* that so impressed Stewart Brand and that became a touchstone of ecological thinking in the seventies. By then, the dominant culture had entered another period of crisis. The New Left, the hippie counterculture, the Black Power movement, and radical feminism all actively rejected the postwar corporate-liberal social order and the assumptions on which it was based. In this context, decentralization again emerged as an explicitly political vision. Bateson's insights into the nature of democracy merged with the sixties' radical critique, and his ecological vision of a system governed by feedback loops provided the counterculture with an alternative model with which to contest the centralization of power in corporate and military hierarchies.[32]

Bateson strove to make his ideas relevant to the new radicalism. In lectures and essays of 1967 and 1968 (later collected in *Steps to an Ecology of Mind*), he contrasted purposive, centrally directed social planning with "wisdom," which he identified with ecological thinking. In a speech delivered at the Congress on the Dialectics of Liberation held in London in July 1967, he proclaimed:

> Purposive consciousness pulls out, from the total mind, sequences which do not have the loop structure which is characteristic of the whole systemic structure. If you follow the "common-sense" dictates of consciousness you become, effectively, greedy and unwise.

In addition to Bateson, speakers at the congress included a pantheon of other cultural radicals of the 1950s and 1960s, including R. D. Laing, Paul Goodman, Stokely Carmichael, Herbert Marcuse, Allen Ginsberg, and Julian Beck. "Lack of systemic wisdom is always punished," Bateson told his younger audience. And yet, he continued—linking his ideas to causes of the late sixties that spanned from antiwar activism to environmentalism—"today the purposes of consciousness are implemented by more and more effective machinery, transportation systems, airplanes, weaponry, medicine, pesticides, and so forth." As a result, "conscious purpose is now empowered to upset the balances of the body, of society, and of the biological world around us."[33] The culture of feedback opposed that trend.

The Culture of Feedback as Practice and Form

The culture of feedback's use of the natural feedback loop as the basis of cultural forms to give new meaning to the idea of freedom is part of the history of improvisation as a method of art-making in America. Models of improvisation from the first half of the century, following templates laid down by surrealism and Jungian psychology, emphasized the revolutionary possibilities of impulses that welled up "from the depths" of the unconscious. However, as Colin Campbell pointed out in *The Romantic Ethic and the Spirit of Modern Consumerism*, imagining such feelings as the site of freedom is not contrary to the consumer capitalist ethos. In consumer capitalism, individual desire, as the motivator of consumption, is the necessary complement to efficient industrial productivity.[34]

By contrast, ecological thinking moves beyond the logic of this opposition. It does not locate freedom either in the maximization of productive capacity or in the maximal fulfillment of individual desire. Instead, it aims for maximum relationality. Correspondingly, the model of improvisation associated with ecological thinking emphasized the emergence of new forms of art through processes of group interaction, or feedback. Through such interactions, the self was regenerated and transformed beginning at it surfaces; and not by impulses emanating outward from an imaginary central core. For example, as Cynthia Novack explained, the dance form from the early seventies known as contact improvisation "tried to shed the concept of . . . the body dominated by an expressive inner self" and to replace it with a focus on "the responsive body."[35] The result was a form that offered its dancers an exercise in the creation of systemic intelligence.

Successful improvisation along these lines was seen as participating in a natural process akin to evolution. Like other versions of evolution in nature, such improvisations made use of indeterminacy, diversity, and the constraints imposed by the dynamics of interaction. The last played a defining role in bringing form to emerge out of the first two. Thus experimental composer Brian Eno stated, "I want to be on the edge between improvisation and collaboration"—where collaboration refers to the constraints imposed by participation in a group dynamic. Similarly, Ian McHarg, a pioneer of ecological design, defined creativity in 1969 as innovation coupled with "responsibility"—a word connoting both responsiveness and accountability. He had, he wrote, "turned to the world at large in order to find laws and *forms of government* that might *work satisfactorily*," and had found them in ecology. "This way has no central authority," he asserted. "Tyranny is rejected because it suppresses the uniqueness of the individual and his freedom." Yet "anarchy is rejected because it replaces creation with randomness." Instead, "poised between these two extremes is the concept of creation, linked to uniqueness, freedom, and . . . responsibility."[36]

In the chapters that follow, I trace the impact of this idea through seventies culture, following a path that is roughly chronological but organized thematically. Chapter 1 describes how systems ecology emerged in the 1950s from the integration of systems theory and ecology. I then summarize the ecological critique of game theory, as both fields claimed the mantle of postwar cybernetics but took it in very different directions. The chapter goes on to recount how the worldview of systems ecology was brought to the attention of millions of Americans in the late sixties and early seventies by the environmental movement. Environmentalists suggested that the collapse of natural systems was imminent if humans did not quickly learn to bring their behaviors within nature's constraints. This stimulated efforts to define sustainable living practices.

Chapter 2 introduces the ideas of general systems theory, which in the late sixties and early seventies clarified the theoretical underpinnings of ecological thinking. General systems theory described how nested open systems governed by feedback loops had adaptive capacities that were tantamount to the workings of a "mind." General systems theory made it reasonable to imagine animals, plants, and other natural systems as intelligent and sentient entities. Among the most significant models of such a mind was the process of coevolution. The theory of coevolution reimagined the workings of natural selection in a way that emphasized decision making rather than competition. Some envisioned the entire ecosystem as a single sentient and intelligent being named "Gaia."

For many Americans, such insights implied a validation of traditional Native American ideas, which posited a spirit immanent in material nature, unlike the scientific discourses of Western modernity. Referred to in environmentalist and New Age discourses alike as a "reverence" toward nature, the traditional Native American viewpoint was equated with an ecological wisdom that Western civilization had sacrificed in its pursuit of a scientific knowledge that promised to predict, manipulate, and otherwise transcend nature. Chapter 3 explores how this set of ideas was promulgated through popular culture, and what its implications were both for non–Native Americans who saw themselves as adopting traditional Native American beliefs, and for American Indians themselves.

Various efforts to communicate with the sentience of plants form the subject of chapter 4. According to systems theory, mental processes are not restricted to conscious thought. The physical body's kinesthetic and sensory pathways are key aspects of intelligence and comprise part of a more inclusive and holistic intelligence than consciousness alone can provide. This holistic intelligence gives rise to an intuitive grasp of the environment that registers as "feelings." Empathy constitutes communication on this level of intelligence. In this light, even plants could be viewed as intelligent, since they respond to their environments via electrical impulses that might be compared to bodily affects in humans. While it was exciting to imagine an almost ubiquitous vegetal intelligence, it was also at the same time unsettling, since a view of plants as sentient subjects implicitly threatens to undo the mystique with which we psychologically surround our own subjectivity.

Certain styles of music appeared to offer a middle ground where two such disparate intelligences as the human and the vegetal might meet. Chapter 5, "Ambient Music," examines how experimental composers explored the possible use of music as a means of engaging in an intersubjective dialogue with sentient nature. They invented various ways of making music that interfaced with nature's systemic intelligence by imitating its processes, using feedback loops to integrate the sounds of daily living into "living" environments of sound.

Chapter 6, "Dancing with Animals," describes how ecological thinking's emphasis on empathic interactions with nature led to a particular excitement about exploring new forms of relatedness to animals. Acknowledging the emotional lives of animals demanded moving beyond behaviorist approaches. This fostered a particular interest in relationships with horses and small toothed whales (dolphins and killer whales), because these were two groups of animals that were known to resist behavioral conditioning. The intended rapport was often described as a kind of dance, due to its reliance on empathy, interaction, and physicality. A range of cultural practices including flotation therapy, "dynamology," contact improvisation, and horse whispering fostered empathic connections between humans and animals through innovative choreographies. At the same time, multiple works, from Jane Goodall's book *In the Shadow of Man* to Godfrey Reggio's film *Koyaanisqatsi*, criticized the choreographies that humans imposed on animals by confining and domesticating them, and encouraged people to see themselves as animals in captivity.

In chapter 7, I analyze how a resurgent conservatism associated with the ascendency of President Ronald Reagan triumphed politically over ecological thinking in the early 1980s. Both orthodox neo-Darwinists and free-market economists rejected ecological thinking in favor of a reinvigorated game theory. In debates surrounding the publication of E. O. Wilson's *Sociobiology*, geneticists mocked the theory of coevolution and the ecological vision of a post-Cartesian science. In a related move, free-market advocates rejected the environmentalist credo that there were natural limits to human population and economic growth. Embracing traditional ideas of independence and an emphasis on defensible perimeters that ran counter to ecological thinking's idea of identity as emergent from the interdependence of multiple layered systems, Reaganism promised to restore to voters "control of their own lives."

In the concluding chapter, I reexamine the most well-known histories of seventies culture in light of this new information. Recovering the history of the culture of feedback compels us to reimagine the seventies as something other than the decade of malaise. Ecological thinking offered a model for different forms of relationship—both among people, and between people and the rest of nature—that continues to shape the American path into postmodernity.

ONE

Systems, Ecology, and Environmentalism

Maturity, sanity, and diversity go together. GARY SNYDER

Initially and most simply in the 1940s and 1950s, the cybernetic understanding of intelligence as a phenomenon of feedback loops referred to the capacity of a system to learn or self-correct. But in the 1960s, systems theorists devoted increasing attention to how such systems were "nested," with feedback loops inserted within feedback loops so that what constituted the "outside" environment of a system on one level was itself subject to the dynamic forces of a larger system of which both were subsystems.[1] The fact that natural systems were typically nested in this way implied that intelligence involved not only the capacity to learn, but a further capacity to respond *creatively* to changing environmental circumstances: to learn, spontaneously, new ways of learning—or, in a word, to evolve. The "new ecology" that emerged through the influence of cybernetics envisioned nature as such an evolving, self-regulating system, governed by feedback loops that placed constraints on the behaviors of its various parts.

The first textbook explicating the systems view of ecology was published by Eugene and Howard Odum in 1953. What begs for historical explanation, therefore, is the suddenness with which ecological thinking gained popular traction in the early 1970s. Paul Brooks, editor in chief at

Houghton Mifflin Publishing Company in Boston, wrote in 1970 that "the science of ecology has quite suddenly emerged from the obscurity of academic studies to become a household word."[2]

This sudden prominence was due largely to the impact of the environmental movement, which brought ecological thinking to the attention of millions of Americans. Environmentalism used ecology to assert that if people did not quickly bring their behaviors within sustainable bounds, the collapse of the natural system supporting them would force a radical and painful remedy upon them. The mass popularity of the first Earth Day in 1970 expanded on the work of environmental writers who had brought the principles of ecology to general attention during the previous decade.

The idea of ecosystem collapse as nature's feedback to errant humans formed the basis of an ecological ethic. Gregory Bateson pronounced in 1970 that "the processes of ecology are not mocked." He was paraphrasing the biblical injunction "God is not mocked; for whatever one sows, that will he also reap." By substituting "the processes of ecology" for "God," Bateson implied that, like the God of the Bible, the forces of ecological feedback both dictated ethical behaviors and meted out punishments for their transgression.[3]

Systems Ecology and Information Theory

In the seventies the idea of the ecosystem as a community of mutual interdependencies offered a significant popular alternative to the two dominant views of nature previously extant in American culture: that of nature as a savage wilderness to be subdued and civilized; and that of nature as a resource given significance only through human utilization. The term "ecosystem" was coined in 1935 by British ecologist Arthur Tansley to refer to the dynamics of natural communities such as were then being described by American ecologists like Aldo Leopold and Paul Sears. It was taken up in earnest in the 1950s and 1960s by a new generation of ecologists feeling the impact of cybernetics, among them Ramón Margalef and the brothers Eugene and Howard Thomas (H. T.) Odum. Yale ecologist George Evelyn Hutchinson, who participated in the Macy conferences on cybernetics, published a paper in 1948 titled "Circular Causal Systems in Ecology," applying cybernetics to the ecological problem of mapping the causes of population changes in biological communities. H. T. Odum was Hutchinson's doctoral student. In his 1950 dissertation, Odum argued that natural selection was best modeled not as a competition among individuals, but as a history of open systems. The community that is successful in maintaining itself, through a combination of homeostasis and evolution, survives. In ecosystems ecology, as Sharon Kingsland explained, the ecosystem as a whole is understood "as a self-regulated entity, like an individual organism."[4]

Called the "new ecology" or "systems ecology," ecosystem ecology focused on the flow of energy and matter through ecosystems, detailing their dynamics and the causes and effects of disruptions in the flow. These processes could be modeled in the language—and sometimes also the mathematics—of feedback loops. Engaging the cultural debate over the meaning of efficiency, in the mid-1950s Odum published an essay in *American Scientist* that used systems of equations to define maximum efficiency as the slowest rate of increase in entropy.[5]

Entropy is the energy that dissipates in the course of an action (Odum called this "leakage"). According to the second law of thermodynamics, in the universe as a whole, entropy can only increase over time. The energy of the sun's rays flying off into space, for example, dissipates for the most part into useless energy as it scatters. The exception, of course, is the portion of the sun's energy that is captured by Earth's plants and sent through photosynthesis into the food chain: the cycles of energy transfer within an ecosystem. This energy dissipates at a much slower rate—depending on the complexity of the community dynamics it gets entered into. Thus life and living systems, as they proliferate, slow the rate at which entropy inevitably increases.

The flow of entropy was most impeded, Odum demonstrated, by the functioning of nested open systems governed by feedback loops. Quantitatively, according to his equations, maximum efficiency as defined in this way never amounts to more than half of the maximum efficiency suggested by the Taylorist ratio of energy output to energy input.[6] But Taylor's method of calculation ignores all sorts of externalities. Odum's method indicates the best system holistically—that is, if externalities are not disregarded. It maximizes the power available to be retrieved and recirculated repeatedly through the system before eventually being lost.

Toward the end of his essay, Odum offered examples to illustrate that his reasoning could be applied to all nested open systems, whether "physical, biochemical, biological, ecological, or social." Thus he expanded ecosystems ecology to include human factors. This would come to be called "human ecology," a subfield that received increasing emphasis during the 1960s.[7]

Significantly, according to Odum's model, diversity is key to creating maximum systemic efficiency, since it provides for a proliferation of different energy pathways. The more various and complex the paths of energy circulation, the more the progress of entropy is slowed. Odum in this way provided a mathematical confirmation of Paul Sears's 1935 assertion that monoculture on the arid plains had hurried the rate of soil erosion by destroying the plant communities that buffered against extremes of temperature and wind.[8]

Another insight of Sears's that Odum's entropic model of efficiency corroborated was that it was an illusion to see technological progress as opening out to infinite future possibilities. In reality, technologies could only give people new ways to tap energy resources stored up by biological communities in the finite past. While they have enabled a progressively more thorough tapping of those stored resources, they have done so at a progressively higher cost in entropy—at lower and lower efficiency rates, from an ecological viewpoint. For Sears, increased rates of soil erosion symbolized this higher cost of doing business. "Fertility has been consumed and soil destroyed at a rate far in excess of the capacity of either man or nature to replace," he warned. Odum phrased the problem in terms of energy dissipation: "The rate of entropy increase in maintaining our American civilization with fuels is very great," he observed. And the Cold War competition between nations to maximize the amount of energy they could afford to expend in fighting one another was not ecologically conducive to the long-term survival of either.[9]

Between 1956 and 1958, Spanish ecologist Ramón Margalef translated Odum's ideas into the language of information theory. One definition of information is that it is structure imposed on variety. (In a binary information-storage system, "1" can mean something different from "0," as long as you can impose those two distinct states.) Completely random variation contains no information. Complete randomness is also the state of total entropy, making it possible to conceive of information as the contrary of entropy, or "negentropy." This in turn makes it possible to see Odum's ecosystems, which circulate energy and prevent its dissipation as entropy, as forms storing useful information. The system with maximum ecological efficiency contains the maximum possible information or, in other words, is the form with the most learning. In his 1968 book, *Perspectives in Ecological Theory*, Margalef wrote:

> The process of [ecological] succession is equivalent to a process of accumulating information. The initial, poorly organized, stages receive the full impact of the environment and any changes in it. Individuals of different species are selectively destroyed. . . . [But] in time the acquired information is expressed in a new organization of the ecosystem. This organization takes into account the predictable changes in the environment, and even controls the environment, so that in the future much smaller changes in the community are necessary to keep it in stable occupation of its area. . . . One can say that the ecosystem has 'learned' the changes in the environment, so that before change takes place, the ecosystem is prepared for it.[10]

By applying the mathematics of information-theory cybernetics to the dynamics of populations in an ecological community, Margalef concluded that the diversity of species in a community was indexical of its health, in terms of how far away in time it was from experiencing a fatal instability. Not only biological diversity was important, but also diversity of interactions. Because all successful adaptive responses are only temporarily advantageous, evolution is understood to take many paths simultaneously. Diversity implies more flexibility and, by extension, improved sustainability.[11]

According to the conclusions that Odum and Margalef reached by applying cybernetics to ecology, the most diverse and complex systems were the most knowledgeable, the richest in internal resources, and the most resilient; hence, the longest persisting.[12] Elimination of the superfluous only took place in immature, poorly organized systems. In mature systems, everything was integrated through feedback loops, so nothing was superfluous and only the minimum was eliminated.

Ecological Thinking versus Game Theory

At the same time that ecosystem ecology was exploring the implications of cybernetics for natural systems and extending its conclusions to human ecology, a competing discourse also related to cybernetics, called "game theory," was being used by military strategists to assist them in Cold War decision making. The adherents of ecological thinking and game theory often found themselves at odds.

Modern game theory was pioneered by émigré mathematician John von Neumann, who regularly participated in the postwar Macy Foundation conferences. Von Neumann's game theory used the mathematics of homeostatic systems that Norbert Wiener had made the basis of cybernetics, and applied it to interactive decision-making scenarios that were called "games." As William Poundstone explained, "A 'game' is a conflict situation where one must make a choice knowing that others are making choices too, and that the outcome of the conflict will be determined in some prescribed way by all the choices made."[13]

Unlike ecosystems ecology, which emphasized the symbiotic functioning of whole systems, game theory focused on competition. Its equations calculated how one could statistically maximize one's competitive advantage over an intelligent rival. When applied in military planning, economics, or evolutionary biology, the usefulness of game theory's mathematics was premised on the existence of a population of completely self-interested players competing logically and yet ruthlessly for scarce resources, whether territory, food, or mates. The classic games on which the field was built were zero-sum games, meaning that one player benefits only when another player loses. The field was extended to include non-zero-sum games and games of more than two players by mathematician John Nash—although in such games the "optimal" strategies, known as "Nash equilibria," often proved disadvantageous to all parties.[14]

Game theory was adopted by American military strategists to model the possible future outcomes of their present decisions, since the unprecedented possibilities of nuclear warfare seemed to make prior military experience irrelevant. In the late 1940s, von Neumann was brought on board as a consultant at the RAND Corporation, a think tank housed at the Douglas Aircraft plant outside Los Angeles and exclusively dedicated to military strategizing in the context of atomic warfare. There, game theory helped to propel the emerging atomic arms race by influencing the Truman administration to develop and stockpile hydrogen bombs. Although Robert Oppenheimer, as head of the General Advisory Committee to the Atomic Energy Commission, counseled President Truman in 1949 to set a good example by shelving the H-bomb project ("In determining not to proceed to develop the Superbomb," Oppenheimer wrote, "we see a unique opportunity of providing by example some limitations on the totality of war"), he was overruled by Secretary of State Dean Acheson, who argued that it was too risky to assume that the USSR would not take advantage of a similar opportunity for nuclear superiority.[15]

Game theory encouraged players to see one another as implacable foes. In a letter to Norbert Wiener in 1952, Gregory Bateson identified the problem as game theory's inability to take deutero-learning into account: "What applications of the theory of games do, is to reinforce the players' acceptance of the . . . premises, and therefore make it more and more difficult for the players to conceive that there might be other ways of meeting and dealing with the other," he wrote. But von Neumann's theories were nevertheless embraced at RAND, especially by the futurist Herman Kahn, who argued in his 1960 book *On Thermonuclear War* that the United States should prepare to wage and win a nuclear war, and not rely merely on deterring one.[16]

That Wiener consulted with Bateson while von Neumann collaborated with Kahn illustrates that, although they shared a mathematics, profound differences of worldview separated the two men. According to historian Steve Heims, von Neumann "would much rather err on the side of mistrust and suspicion than be caught in wishful thinking about the nature of people and societies." Wiener, on the contrary, believed that game theory overstated the role of malevolent competitiveness in decision making. In *Cybernetics,* Wiener derided "von Neumann's picture" as not being a true systems theory, but a distorted abstraction that amounted to "a perversion of facts." In his 1950 treatise, *The Human Use of Human Beings,* Wiener made the ecological argument that in systems, the "enemy" was entropy—"the absence of order"—rather than the actively malevolent "contrary force" that von Neumann's game theory posited.[17]

From an ecological perspective, game theory falsified the actual conditions of decision making in several ways. For one, it depended on a sequential logic, in which one player decides a move, then another player decides a move, and so on—what neurobiologist Paul Weiss called a *"linear cause effect cause effect . . . chain causality."* It was therefore well suited to computers, which operated on that basis. But it did not proceed in a manner true to life, in which decisions are often made simultaneously and in effect continuously. The organism does not process all inputs equally and sequentially, but perceives selectively, its sensorium responding to unconscious prioritizations of different needs at different times and in different environmental contexts. "The dynamics of living systems are not those of linear arrays," Weiss asserted; and their lines of feedback not only branch, but are cross-linked into "a web of great complexity"[18] (see fig. 3).

Related to this critique was the problem that making game theory into decision theory—that is, applying it to real-life situations—required one to first construct a hierarchical list of values that the players would strive to maximize. In real life, values and priorities were seldom easily identifiable or permanently fixed, as they were in a game. Moreover, as Bateson pointed out to Wiener, game theory left no room for deutero-learning; whereas, in actual complex systems,

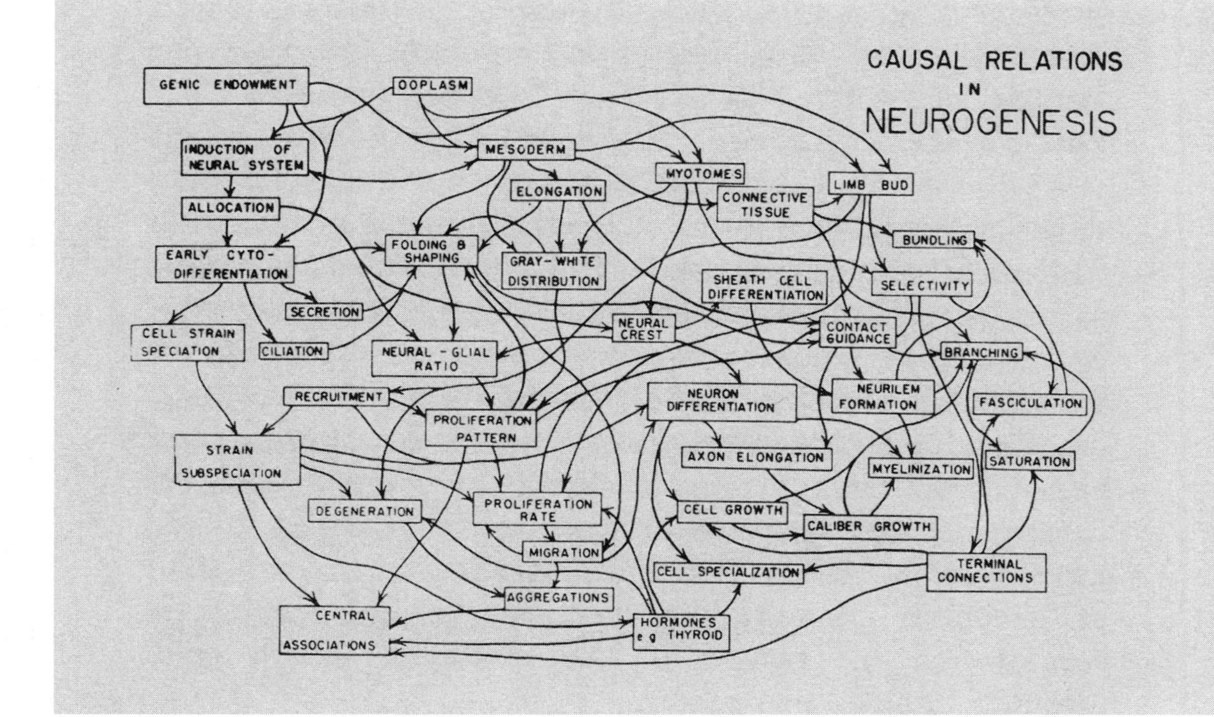

Figure 3. Complex causality, from Paul Weiss's "The System of Nature and the Nature of Systems" (1977).

multiple levels of decision making and valuation (including learning and deutero-learning) could affect one another in ways that sequential logic could not map without arriving at paradox.[19]

Finally, for games with incomplete information (literally all strategic games), game theory relies on statistical averages to assess the success of a strategy. This means, as Heims observed, that "a sound policy can be recommended to each player who plays many rounds of the game, but no best strategy is available for playing only one round." In real-world decision making, however, there is typically only one round. Actions take place in an irreversible flow of time—Wiener referred to it as "Bergsonian time" as contrasted to "Newtonian"—in which they create new conditions for the system, which cannot then be reset back to its previous state. One implication of this is that, in life, means cannot be neatly separated from ends. As Heims wrote, "In game theory ends and means are neatly separated: the 'outcomes' are judged by their desirability, while the 'strategies' are judged solely by their efficaciousness in bringing about the desired ends."[20] But events in reality have no clear finish line at which the winner will be declared; at each juncture, new challenges require creative adaptions, and yesterday's strategies (with both their intended and unintended consequences) furnish the only available basis for today's options.

Wiener therefore exhorted that "the true condition of human life" be seen as an "irreversible movement into a contingent future," demanding for its successful negotiation the "social feedback" of "democracy." Game theory, on the contrary, he accused, demanded secrecy and fostered stagnation. Von Neumann, like Kahn, favored economic expansionism and the centralization of power as means for America to maintain its global supremacy. For Wiener, democracy and decentralization were America's primary cultural assets. He worried in 1948 that his contributions to cybernetics would foster the centralization of power. He was also wary of the economics of corporate liberalism, criticizing in *Cybernetics* the myth that the "free market" was a "homeostatic process" that could be relied on to maintain the health of the "body politic." "Unfortunately, the evidence, such as it is, is against this simple-minded theory," he wrote. "The market is a game."[21]

In *The Human Use of Human Beings*, Wiener expanded this caveat into a broader cultural critique that resonated with the earlier writings of Dust Bowl ecologist Paul Sears. "To the average American, progress means the winning of the West. It means the economic anarchy of the frontier," Wiener began. "What many of us fail to realize is that the last four hundred years are a highly special period in the history of the world," set in motion by the discovery of the Americas with resources that "seemed inexhaustible" and "encouraged an attitude not unlike that of Alice's Mad Tea Party." Yet in the recent past, he warned, "the rate at which one seat has been abandoned for the next has been increasing at [an] . . . increasing pace." In language that he later omitted from the revised second edition of the book, Wiener asserted that

> so long as anything remained of the rich endowment of nature with which we started, our national hero [in the U.S.] has been the exploiter who has done the most to turn this endowment into ready cash. In our theories of free enterprise, we have exalted him as if he had been the creator of the riches which he has stolen and squandered.

These observations led him to invert the common wisdom regarding modernity:

> In the opinion of the average man, the modern period is characterized by what he considers to be the virtuous rapidity of progress. It will be quite as true to say that the modern period is the age of a consistent and unrestrained exploitation: of an exploitation of natural resources; of an exploitation of conquered so-called primitive peoples; and finally, of a systematic exploitation of the average man.[22]

From this reference to "conquered so-called primitive peoples" and "exploitation of the average man," Wiener's critique segued into a rejection of racism and Taylorism. From a systems theory viewpoint, he explained, diversity is necessary in that it enables a system "to modify [its] pattern of behavior into one which in some sense or other will deal more effectively with the future environment . . . and its future contingencies." Therefore the doctrine of white supremacy was socially deleterious. Yet even the partial democracy that existed in racist 1950s America was "too anarchic for many of those who make efficiency their first ideal. These worshippers of efficiency would have each man . . . perform a function to which he is bound." From an ecological standpoint, this was not efficiency but waste: "Variety and possibility belong to the very structure of the human organism," Wiener wrote; and rigid social structures "throw away this enormous advantage that we have"—they "throw away nearly all our human possibilities and by limiting the modes in which we may adapt ourselves to future contingencies, they reduce our chances [as a society] for a reasonably long existence on this earth."[23]

Likewise, game theory, due to the rigidity of its premises, was endangering our future. Used in war, Wiener warned, it "will in general lead to an indecisive action which will often be not much better than a defeat." Wiener died in 1964; but the pushback against game theory that he for a time spearheaded gained momentum during the radical sixties. Radical economist Gunnar Myrdal spoke out at Harvard in 1971 against the fashion of using "utterly simplified models [of social reality] . . . given mathematical dressing."[24] But the most prominent public critic of game theory that year was Daniel Ellsberg, who smuggled thousands of pages of classified documents out of the Pentagon and into the hands of journalists in order to expose the shortcomings of American policy in the Vietnam War. In the face-off between ecological thinking and game theory, Ellsberg had switched sides. A game theorist and strategic analyst for the RAND Corporation, while still in his twenties he had helped Kahn write *On Thermonuclear War.* As late as 1962, he shared Kahn's hawkish views.[25] But over the next decade, he changed.

In the spring of 1971, only weeks before the *New York Times* began to publish the Pentagon Papers, Ellsberg authored an article in the journal *Public Policy.* In it he provided a detailed critique of American decision making in Vietnam, beginning with how President John F. Kennedy, influenced by game theory, had laid the foundations of the future war. Ellsberg rejected the metaphor of Vietnam as a "quagmire" that America had "stumbled" into with "good intentions." A more apt metaphor, he wrote, would be "The Stalemate Machine." In an expanded version of the article published in 1972 (while he was awaiting trial for espionage), he described how the "decision rules" of game theory had fostered military escalation, a dependence on the government of South Vietnam, and covert operations, to make the American war effort a no-win/no-lose stalemate. Given the impossibility of achieving the ostensible long-term goal—"the assurance indefinitely of a totally non-Communist regime" in a democratic Vietnam where Communism enjoyed broad popular support, American leaders had chosen to "concentrat[e] almost exclusively upon the aim of minimizing the short-run risk of anti-Communist collapse." Decisions were therefore optimized for their value as "holding actions, adequate to avoid defeat in the short run." "Rule 1 of that game is: *Do not lose the rest of Vietnam to Communist control before the next election.*" For this reason Ellsberg rejected the quagmire metaphor as a self-serving obfuscation on the part of the decision makers: since the "decisions actually were [consciously oriented], when chosen, to the defensive aim of averting an immediate

Communist takeover, *each of these decisions of the past two decades can be said to have achieved its initial, internal aim*."[26]

Ellsberg joined the antiwar movement because of his gradual disillusionment with the American political and military establishment. But other influences were also simultaneously at work. One of these was his respect for the poet Gary Snyder. In January 1960, having enthusiastically read Jack Kerouac's 1958 roman à clef *Dharma Bums*, which featured Snyder as the character Japhy Ryder, Ellsberg determined to visit the Ryōan-ji Zen temple in Kyoto while he was on assignment in Tokyo for the RAND Corporation. That evening, he ran into Snyder in Kyoto by coincidence, and they spent the night arguing the relative merits of pacifism and belligerence. Although Ellsberg was not convinced by Snyder's pacifist stance, he wrote later that he recognized Snyder then as "a better man than I was. . . . I was as smart as he was, but he was wise." Ellsberg felt that his own role, helping the government lie to the public about its military operations, was "not Right Livelihood," as Snyder understood it; and, he continued, "my memory of him that weekend stayed with me in the back of my mind as a kind of touchstone: an image of an alternative way of living." Ellsberg partially attributed to this image of Snyder his later decision "to speak truth to the world," referring to his release of the Pentagon Papers as "doing it [Snyder's] way." In September 1970, on the run from the FBI and carrying the still-unpublished Pentagon Papers in the trunk of his rental car, Ellsberg sought out Snyder again as one of "just a few people I especially wanted to see" before he was arrested. For a second time by coincidence he found Snyder, this time camped in a forest outside of Nevada City, California. "I didn't show him any papers from the trunk so as not to implicate him," Ellsberg wrote, "but I hinted that he was implicated anyway, in the process of my awakening. I wanted to thank him."[27]

Gary Snyder's Ecological Ethic

Associated with both the Beat generation of the 1950s and the environmental movement of the 1970s, the life and work of Gary Snyder epitomize the cultural continuities linking the culture of feedback to the postwar "culture of spontaneity." Beginning in 1947, Snyder had studied Native American cultures and literature at Reed College in Portland, Oregon. In 1952 he abandoned graduate work in linguistics and anthropology to focus on writing poetry, and he began learning Chinese and Japanese in preparation for studying Zen Buddhism in Japan. He helped to organize the famous poetry reading at the Six Gallery in San Francisco in 1955, where Allen Ginsberg gave the first public reading of "Howl." The next year, he left California to study Zen in Kyoto and spent most of the next twelve years in Japan.

As the radicalism of the sixties heated up, Snyder like Bateson found his ideas drawn into political relevance. He read his poetry to the crowd at the first Human Be-In at Golden Gate Park in 1967 and spoke during the protest at People's Park in Berkeley in 1969. Profiled in *Ecotactics*, the Sierra Club's manual for Earth Day activists published in 1970, he emphasized the affinities between ecological thinking and Zen Buddhism: "Ecology demonstrates on the empirical level the myriad interrelationships in nature," he is quoted as saying. "Buddhism, on another level, asserts the same interdependence."[28] In 1975 he won the Pulitzer Prize for Poetry for his book *Turtle Island*, which included the environmental essay "Four Changes."

In lectures and interviews, Snyder contrasted two lifestyles or socioeconomic systems. He called these "ecosystem" culture, which lived within the constraints imposed by the local environment; and "biosphere" culture, which lived by exploiting outside resources through such means as colonialism, slavery, and the burning of fossil fuels.

Biosphere cultures, he wrote, have predominated since the times of the Roman Empire, based on their successful exploitation of distant biological systems. This has fostered an ever-more-extensive centralization of power. Western civilization, conceived in this context, developed corresponding cultural pathologies: "All that wealth and power pouring into a few centers had bizarre results," Snyder wrote. "Philosophies and religions based on fascinations with society, hierarchy, manipulation, and the 'absolute.' A great edifice called 'the state' and the symbols of central power."[29]

In the specifically American context, nineteenth-century agriculture, based on the massive expropriation of African and African American labor, accustomed most Americans to a "standard of living that's out of proportion to who and where you are." In the twentieth century, petrochemicals—which produced fossil fuels, artificial fertilizers, and pesticides—furnished a new outside energy source. As a result, Snyder told an interviewer in 1977,

> We live in a nation of fossil fuel junkies, very sweet people and the best hearts in the world. But nonetheless fossil fuel junkies of tremendous mobility zapping back

and forth, who are still caught on the myth of the frontier, the myth of boundless resources and a vision of perpetual materialistic growth. Now that is all very bad metaphysics, a metaphysics that is leading us to ruin.

The difficult-to-accept solution to the impending environmental crisis, he maintained, was a radical return to ecosystem culture: to the "reciprocal and interacting" constraints of "place" and of "the energy-pathways that sustain life."[30] The concrete practice of learning to live in such an ecosystem culture was, he liked to say, "the real work."

In accomplishing the real work, Snyder relied on two different groups of people, as he explained to an interviewer from a local countercultural newspaper: his network and his community. The former were geographically extensive and organized around a common outlook or pursuit, like poetry or yoga or environmentalism. Community, on the other hand, was local, and its members diverse in needs and outlooks. Community, Snyder insisted, grounds ideas and actions in the physical and social realities of place. The local ecosystem brings all its inhabitants together to forge a shared living environment. "You don't all agree with each other and there are problems that you have to live with and work out over a long scale of time."[31]

Most generally, the ecological dynamics of community were encapsulated, for Snyder, in the metaphor of eating. Eating, he explained in a talk in 1974, is the means of cyclical energy transfer within an ecosystem. It passes energy from one organism or subsystem to another, in a "network of gift-exchange." As he told an interviewer in 1976, "The *real work* is eating each other, I suppose."[32]

The Subversive Science

In the late sixties and early seventies, ecological thinking came to prominence in American popular culture through the influence of environmentalism. Environmentalism sprang from human ecology, a subfield that emerged during the sixties from systems ecology's increasing focus on studying human disruptions to natural systems. Paul Sears, author of the 1935 ecological classic *Deserts on the March,* gained renewed fame in 1964 for naming ecology the "subversive science" because it contradicted many truisms of American corporate-liberal ideology. In the same year, Frank Egler attached a similarly political inflection to ecology—which had for the past two decades presented itself as a relatively value-neutral scientific discourse—when he introduced a pronounced ethical tone in his article in *American Scientist* on the effects of pesticides on the ecosystem. "We have acted with remarkable *arrogance* to the whole-nature of which we are a part," Egler insisted. He lamented that while ecology was the science perhaps most necessary to humanity's long-term survival, it was also the one least understood by the general public.[33] That was about to change.

The new popular and political face of ecology that emerged in the late 1960s is apparent from two edited collections published around the end of that decade: Robert Disch's *The Ecological Conscience: Values for Survival* and Paul Shepard and Daniel McKinley's *The Subversive Science: Essays towards an Ecology of Man.* Disch's anthology opens with a poem by Allen Ginsberg and includes essays by other icons of the sixties counterculture, including Gary Snyder, Alan Watts, R. Buckminster Fuller, Paul Goodman, and Thomas Merton. Their writings accompany essays by ecologists, some of whom were rapidly achieving the status of public intellectuals: Aldo Leopold, Barry Commoner, Paul Shepard, and Ian McHarg. In the preface, Disch emphasized "the *radical* implications of ecological knowledge," informing his readers that "the industrialized nations will have to undergo a profound change in the values and attitudes they hold" in order to prevent global catastrophe. He hoped, he concluded, that his book would foster a productive conversation among social scientists, government technocrats, and counterculture utopians to bring about this change.[34]

The Subversive Science was structured along the same lines and with the same intentions. It included writings by Watts, Leopold, and McHarg, as well as historian Lynn White and Native American advocate John Collier. These were interleaved with the work of ecologists G. Evelyn Hutchinson, Paul Sears, Vero Wynne-Edwards, and Frank Egler. In a foreword, Aldo Leopold's son A. Starker Leopold questioned the prevailing notion of progress.[35]

Public awareness of the environmentalist message received an enormous boost from the first Earth Day, celebrated in the spring of 1970. The first Earth Day was to environmentalism what the Montgomery Bus Boycott was to the civil rights movement: the beginning of a mass movement culture. Up until then, the only self-identified "environmental" organization in the country was the Environmental Defense Fund, which historian Adam Rome characterized as a "handful of lawyers and scientists." Although Earth Day is now officially observed on April 22, the first Earth Day was more of an Earth Week, encompassing a network of events involving twenty million Americans, fifteen hundred colleges and universities, and ten thousand elementary and secondary schools.[36] Its name was not officially Earth Day but the First National Environmental Teach-In. It was modeled on the New Left's anti–Vietnam War teach-ins, but with an environmental focus.

The success of the First National Environmental Teach-In energized and brought together activists who had previously felt themselves to have little in common with one another. These included century-old groups focused on wildlife conservation and wilderness preservation; anti-smog campaigners in big cities like Los Angeles and New York; and small-town and suburban anti-litter advocates. Michael McCloskey, who served as the Executive Director of the venerable Sierra Club from 1969 to 1985, described how Earth Day acted as a catalyst for that organization's new environmentalist orientation: "We once had been solely focused on conserving nature, now we were also concerned with pollution, public health, population growth, land use, energy, transportation policy, and almost any other issue touching even remotely upon the environment."[37]

Earth Day registered the popular impact of a political momentum built up over the past decade by scientists and elected officials who had become increasingly concerned about environmental issues. Democratic senators Gaylord Nelson of Wisconsin and Edmund Muskie of Maine had established national reputations in the 1960s by championing environmental legislation to address growing public concerns about quality-of-life issues. Nineteen seventy was a watershed year for environmental causes politically as well as culturally. It saw passage of an expansive new federal Clean Air Act and the establishment of a federal Environmental Protection Agency.[38] The Clean Air Act, which initiated the enforcement of National Ambient Air Quality Standards, is sometimes called the "Muskie Act" because of the senator's central role in drafting the bill. Similarly, it was Senator Nelson's initiative that was behind the organizing for Earth Day that began in September 1969.

Liberal politicians' interest in environmental issues can be traced back to economist John Kenneth Galbraith's 1958 critique of postwar priorities, *The Affluent Society*. In that book, Galbraith urged America's leaders to focus more on the quality of public spaces and public services, rather than on facilitating the accumulation of private wealth. Following Galbraith's lead, in 1963 Stewart Udall, serving as secretary of the interior under President John F. Kennedy, published a book titled *The Quiet Crisis*, which advocated the federal regulation of polluters and the signing of international treaties to protect the atmosphere and oceans. Both the title of Udall's book and its introduction (contributed by President Kennedy) drew parallels between the urgency of the environmental crisis and the Cuban missile crisis of 1962.[39]

"Blight" was what liberal politicians named the deterioration of common amenities that they were combating. It was a word that comprehended issues as diverse as poverty, landscape aesthetics, and environmental damage. In *The Quiet Crisis*, Udall insisted on the importance of complementing the space age emphasis of Kennedy's New Frontier with a renewed attention to the Earth:

> History tells us that earlier civilizations have declined because they did not learn to live in harmony with the land. Our successes in space and our triumphs of technology hold a hidden danger: as modern man increasingly arrogates to himself dominion over the physical environment, there is risk that his false pride will cause him to take the resources of the earth for granted—and to lose all reverence for the land.

In this short passage, Udall invokes several structural binaries that were coming to shape the emerging environmental discourse: technology versus the land; dominion versus harmony; and arrogance versus reverence. Echoing Galbraith's critique in *The Affluent Society*, Udall exhorted, "America today stands poised on a pinnacle of wealth and power, yet we live in a land of vanishing beauty, of increasing ugliness, of shrinking open space, and of an overall environment that is diminished daily by pollution and noise and blight."[40]

Published the year before Udall's *The Quiet Crisis*, biologist Rachel Carson's *Silent Spring* proved even more influential. Carson advocated elevating environmental issues above all others on the liberal agenda, naming "the contamination of man's total environment" as "the central problem of our age." Expanding on a fear of radioactive fallout that had become widespread during the "Ban the Bomb" campaigns of the late 1950s and early 1960s, Carson focused her environmentalist critique on the poisonous effects of chemical pesticides, especially DDT. "In this now universal contamination of the environment, chemicals are the sinister and little-recognized partners of radiation in changing the very nature of the world—the very nature of its life," she warned. Evoking half-forgotten nightmares of the chemical weapons used by the Nazis in the Holocaust, she decried their repurposing in "Man's war against nature."[41]

Carson's argument mobilized a set of structural dichotomies similar to Udall's: technology versus the Earth's systems; the conquest and destruction of nature versus harmony and health. But unlike Udall, she insisted that there was an important distinction to be made between landscape beautification and environmentalism. Manicured lawns and flower beds might appear healthy, but in them lurked an invisible "blight" of illness and death brought by chemical poisons. To make this argument, she invoked ecological theory—the most important difference separating her environmentalism from Udall's conservationism. Because of the complex dynamics relating living things and their environments, she explained, chemical pollution initiated a "chain of evil" leading from groundwater to plankton and then to fish; a "chain of poisoning" from the soil to fungal networks to earthworms and then to birds.[42]

Carson's work proved enormously influential with the public and won several awards. In addition, it spurred other biologists familiar with ecological thinking to venture out of the ivory tower and publicly embrace environmental activism. Two of these, Barry Commoner and Paul Ehrlich, followed Carson's example to become best-selling authors and public intellectuals.[43]

Like Carson, Commoner stressed the need to look beyond the typical strategies of nature conservation and beautification. His book *Science and Survival*, published in 1966, targeted several chemical pollutants that Carson had not covered, including phosphates from detergents and fertilizer runoff, mercury, chlorines, and atmospheric carbon dioxide. But his strongest polemics were reserved for the environmental effects of a possible nuclear war. The newness of his contribution lay in applying ecological thinking to that scenario in order to refute Herman Kahn's vision of a winnable thermonuclear war. "The biosphere is a marvelously intricate system," he explained. "All the parts of this system are mutually interdependent." Therefore the deforestation, radiation poisoning, and nuclear winter that would come in the aftermath of nuclear war would devastate the biosphere and make agriculture impossible, dooming most surviving humans to starvation.[44]

In his book *The Population Bomb*, published in 1968, Paul Ehrlich reframed Carson's critique of industrial agriculture as part of a larger picture of the environmental degradation caused by human overpopulation. Ehrlich introduced his readers to the term "ecosystem" and explained how overpopulation damaged the ecosystem by reducing its complexity, making it vulnerable to instability. Ehrlich joined the Earth Day steering committee at Senator Nelson's invitation in November 1969.[45]

Earth Day marked a cultural shift toward the wider influence of ecological thinking, as environmentalism gained the attention of millions of Americans. Sierra Club membership soared, accompanied by a change in institutional strategy in support of environmental activism. Even the Sierra Club's publications changed: from lavishly illustrated, oversize coffee table books to inexpensively produced "battle books" that offered synopses of environmental issues and arguments useful in convincing deniers.[46] *Ecotactics: The Sierra Club Handbook for Environmental Activists* was published in 1970 especially for Earth Day distribution.

Like Earth Day activities generally, the essays in *Ecotactics* offered the public a crash course in ecological consciousness. In the foreword, Michael McCloskey proclaimed: "A revolution is truly needed—in our values, outlook and economic organization. For the crisis of our environment stems from a legacy of economic and technical premises which have been pursued in the absence of ecological knowledge." The other essays in the pocket-size publication followed suit, repeatedly linking the ideas of ecology to the call for a more comprehensive cultural revolution.[47]

Nature's Feedback

Systems ecology emphasized that nature is a self-regulating system governed by feedback loops. Environmental activists hammered home the notion that the current environmental crisis was the result of such feedback loops responding to human behaviors. Because an understanding of ecological dynamics was needed both to perceive the environmental crisis and to figure out how to respond to it, environmentalist writers persistently returned to the function of feedback loops as constraints in natural systems. In *Ecotactics,* Ralph Nader wrote (echoing Leopold and Sears) that our desire for plenty and control had become excessive, and as a result humanity was encountering negative feedback from nature. "For centuries, man's efforts to control nature brought increasing security from trauma and disease. Cultures grew rapidly by harnessing the forces of nature to work and produce for proliferating populations," Nader recounted. "But in recent decades, the imbalanced application of man's energies to the land, water, and air has abused these resources to a point where nature is turning on its abusers." Rachel Carson similarly titled one of the chapters in her book *Silent Spring* "Nature Fights Back." In it she described how the "complex biological systems" in which pesticides intervene are "living communities" that reestablish a new equilibrium incorporating the chemicals' presence, leading to the development of new chemical-resistant strains of pests. Thus, "chemical controls are self-defeating."[48]

Like many other ecological thinkers, Carson went on to contrast two approaches to efficiency: the engineering and the ecological. The former was ultimately doomed to failure, she asserted, because it relied on simplistic notions of direct cause and effect that did not take feedback dynamics into account. "Single-crop farming does not take advantage of the principles by which nature works," she wrote. "It is agriculture as an engineer might conceive it to be." Real efficiency lay in making use of the natural constraints that are already operative in a complex self-regulating system. "Nature has introduced great variety into the landscape, but man has displayed a passion for simplifying it," Carson wrote. "Thus he undoes the built-in checks and balances" by which biodiversity increases stability. To restore those balances, humans would have to accept constraints on their own behaviors.[49]

One of the most important ways of formulating the shift in thinking demanded by environmentalism was to say that economics must become a subfield of ecology. Barry Commoner invoked this paradigm in a follow-up work to *Science and Survival* published in 1971, titled *The Closing Circle: Nature, Man, and Technology*. In that book he denounced the wealth produced by modern technology as "an insidious fraud," in that its "short-term exploitation of the environmental system" had racked up "a debt to nature . . . so large and so pervasive that . . . the bubble is about to burst."[50]

The effort to reformulate economic principles along the lines of ecological thinking was initiated by economist Kenneth Boulding in an influential essay published in 1966 called "The Economics of the Coming Spaceship Earth." Adopting the systems ecology of the Odums and Margalef, which redefined efficiency in terms of maximizing the energy recycled within the system, Boulding argued for replacing what he called the "cowboy economy"—a historical legacy of the American frontier ethos—in favor of a "spaceship economy" that acknowledged environmental constraints on economic expansion. The cowboy, Boulding wrote, "being symbolic of the illimitable plains and also associated with reckless, exploitative, romantic, and violent behavior," symbolized a pervasive but outmoded attitude: one that celebrated the subjugation of nature to achieve increased productivity. Cowboy economics was embedded even in our measures of economic health, such as the gross national product, which equates growth with increased production and consumption. From an ecological standpoint, to the contrary,

> the essential measure of the success of the economy is not production and consumption at all, but the nature, extent, quality, and complexity of the total capital stock, including in this the state of the human bodies and minds included in the system. . . . [T]he less consumption we can maintain a given state with, the better off we are.

Economic health could only be understood in terms of a feedback loop, not in terms of linear growth, Boulding wrote, since economically "man must find his place in a cyclical ecological system."[51]

Boulding's analysis was recapped for Earth Day participants by John Mitchell (the editor in chief of the Sierra Club) in the book *Ecotactics*. Mitchell drew out the cultural implications of Boulding's critique, drawing a parallel between Boulding's "cowboy" and the modern-day engineers that Rachel Carson criticized for being too intent on devising technological fixes for ecological challenges. Mitchell criticized the "cowboy-technocrat" for adhering to a simplistic worldview that failed to acknowledge natural feedback loops and therefore caused unanticipated disasters. "Behind every ecological boomerang lurks the cowboy-technocrat," he accused, who was unable to see the unintended long-term effects of his interventions. What was needed to address the environmental crisis was not more of the same, but a new consciousness respecting nature's own feedback mechanisms.[52]

Like the "anti-Western" movies of the time—Arthur Penn's *Little Big Man* (1970) and Robert Altman's *McCabe & Mrs. Miller* (1971) are among the best known—Boulding and Mitchell recast the myth of America's westward expansion as a story of failure. Rather than a social Darwinist "winning of the West"—the triumphant march of a superior civilization—they presented it as driven by an inability to live sustainably on the land. "The American pioneer . . . clear-cut the forest," Mitchell recounted. "Next he planted his crops. Then he failed to understand why the land went stale with erosion. So he moved on, beyond the Western hills, and cut again and planted his fields and once more failed to understand."[53]

Maverick designer R. Buckminster Fuller adopted and expanded on Boulding's idea of the spaceship economy in his *Operating Manual for Spaceship Earth*, published in 1968. Fuller admonished, "Up to now we have been misusing, abusing, and polluting this extraordinary chemical energy-interchanging system for successfully regenerating all of life aboard our planetary spaceship." Now, however, in order to "avoid extinction as unfit," humans had to conceive of the interrelated "energy patternings" of that system "holistically" and adapt their living strategies accordingly.[54]

To adopt the ecological viewpoint meant accepting ecological constraints as ethical guidelines. "Let us accept the proposition," Ian McHarg, a pioneer of ecological design, wrote in his 1969 book *Design with Nature*, "that nature is process, that it is interacting, that it responds to laws, representing values and opportunities for human use *with certain limitations and even prohibitions to certain of these*." Then, the destruction of beach dunes and the grasses that anchored them, for example, would become an ethically prohibited behavior.[55]

Rural communes dedicated to living within the constraints of the local ecosystem and minimizing the increase of entropy inherited much of the momentum of the sixties counterculture. The premise of Stewart Brand's *Whole Earth Catalog* was that they would need both ideas and appropriate technology in order to succeed. Buckminster Fuller's ideas were featured regularly in the *Catalog*, the first section of which was titled "Understanding Whole Systems." Other journals also sprang up to provide ideas and encouragement, including *Countryside* magazine in 1969 and *Acres U.S.A.* and *Mother Earth News* in 1970. These placed less emphasis on theoretical abstractions like cybernetics, and more on the practicalities of how to survive using do-it-yourself renewable energy sources and organic farming.

In his book *The Reenchantment of the World*, social critic Morris Berman argued that ecological thinking offered a moral compass at a time when the radicalisms of the sixties were in need of direction yet radical cultural change was still necessary. Ecological thinking's literal groundedness in natural interactions seemed to offer a necessary reality principle through which to filter the purported discoveries of a psychedelics-infused cultural revolution that had proven susceptible to delusions like the cult of Charles Manson. "In a world that is rooted in bioregional realities," Berman asserted,

> deprogramming does not lead to . . . adoration of the shaman, but of the mystery he makes manifest: the God within, and the ecosystem that reflects it. . . . In short, it is my guess that preservation of this planet may be the best guideline for *all* our politics, the best context for *all* our encounters with Mind or Being . . . the ultimate safety valve in the emergence of a new consciousness.[56]

Significant experiments in sustainable living were undertaken at Synergia Ranch south of Santa Fe and at the New Alchemy Institute near Woods Hole, Massachusetts. John Allen and Marie Harding founded Synergia in 1969, dedicated to rehabilitating exhausted ranchlands. Will McLarney, John Todd, and Nancy Jack Todd established the New Alchemy Institute in 1970; it focused on developing do-it-yourself technologies for aquaculture and hydroponics, renewable energy sources, and small-scale organic farming. Stewart Brand visited New Alchemy in 1973 and awarded it $16,000 in grant money from a fund created by his liquidation of the *Whole Earth Catalog*.[57]

Gary Snyder's ideas were prominently featured in publications aimed at sustainable living. The first issue of *Mother Earth News* led off with an environmental manifesto of Snyder's that the editors had lifted from the pages of the *Whole Earth Catalog* and that was later included in his book *Turtle Island*. Snyder wrote in that essay, titled "Four Changes," that "as the most highly developed tool-using animal, we must recognize that the unknown evolutionary destinies of other life forms are to be respected, and we must act as gentle steward of the earth's community of being." Elsewhere, citing the works of Wendell Berry, Snyder reiterated his support for experiments in living near to the earth in small cooperative communities, redeeming lands that had been exploited and abandoned by the lumber industry and agribusiness. To accomplish this, he told an interviewer in 1971, would require appropriate technologies, both old and new: "There's a lost technology [of the Native Americans], and there's a ghost technology that was never developed" of "decentralization and decentralized energy sources"—for example, of "small-scale decentralized hydro-electric plants" that could power a family ranch from the water in its own creek.[58]

Sustainable-living communities experimented with using new engineering technologies as a means to sustainability. But there is an important distinction to be made here among the varieties of technological utopianism. The attitude prevailing in ecological circles was not excitement over the technological mastery of nature and a transcendence of its constraints, but humility in the face of nature's complexity and a hope to live within its constraints. This disposition is evident in the repeated recourse to terms like "reverence" and "veneration" in describing the proper relations of humans toward nature. At the end of his 1972 book *Introduction to Systems Philosophy*, for example, Ervin Laszlo derived an ethics from his metaphysics that he summarized as "*reverence for natural systems*." He contrasted this to a prevailing attitude of "exploitation" and "arrogance" toward nature that was informed by a "mechanistic" and "manipulative" mind-set. Likewise Arne Næss, articulating the ethical principles of what he called "deep ecology" in 1973, emphasized a "*veneration* for ways and forms of life." Ecological designer Ian McHarg counseled his readers, "Look to the plants, say, 'Through you we breathe, through you we eat, through you we live.'"[59]

TWO

Self-Organizing Systems and Mind in Nature

. . . one ecosystem
in diversity
under the sun
With joyful interpenetration for all.
GARY SNYDER, "FOR ALL"

In the early 1940s, when Aldo Leopold titled his pioneering ecological essay "Thinking Like a Mountain," he used the term "thinking" poetically to refer to the intelligence immanent in natural systems; in the seventies, however, it came to be meant literally. Correspondingly, when Norbert Wiener coined the term "cybernetics" in 1948 to describe how the dynamics of homeostatic mechanisms invest machines with goal-directed behavior or "purpose," he balked at the terms "mind" and "intelligence." But British psychiatrist W. Ross Ashby became interested in Wiener's mathematics and in his 1952 book *Design for a Brain* suggested it could be applied in a description of human neural activity.[1] Neurobiologist Paul Weiss then argued that mind consists of more than a sequence of neurons firing signals across synapses: it also includes the ways in which these very neural pathways develop —what is now called neuroplasticity (see fig. 3).

Weiss, Wiener, and Gregory Bateson shared a common conviction that intelligence must include what Bateson called deutero-learning: the process of learning new ways to learn. Wiener dreamed of a future cybernetics in which machines were self-organizing systems capable of evolving in that manner—what another participant in the Macy conferences, Heinz von Foerster, named "autopoiesis." As Wiener wrote in *The Human Use of Human Beings*, this would be not just a purposeful machine "but a machine . . . [that] will look for purposes which it can fulfill."[2]

Howard T. Odum and Ramón Margalef used systems ecology to suggest that such systems were already operant in nature generally. This made it easier to imagine that ecosystems and other nonhuman natural systems had "minds" similar to the plastic neural networks of humans. Breaking down the mystique with which the Cartesian dualism had surrounded mind and consciousness, in order to see them as emergent phenomena produced by physical feedback loops, was a major factor in the subsequent development of the culture of feedback.

The task of expanding systems theory into a philosophy of how mind was immanent in matter was taken on by thinkers who called their field "general systems theory." The most influential writers of general systems theory were Ervin Laszlo, Arthur Koestler, Gregory Bateson, and James Lovelock. Bateson told his listeners in the late 1960s, "In World War II it was discovered what sort of complexity entails mind. And, since that discovery, we know that: wherever in the Universe we encounter that sort of complexity, we are dealing with mental phenomena. It's as materialistic as that." There was therefore a larger mind "of which the individual mind is only a subsystem . . . [one] immanent in the total interconnected social system and planetary ecology."[3]

The biological theory of coevolution, which described natural evolution as the result of networked subsystems engaged in decision making in concert, lent scientific validity to the conclusions of general systems theorists. As opposed to the orthodox Darwinian model of evolution, the theory of coevolution minimized the role of chance in adaptation, seeing evolution as a process of natural systems engaged in complex loops of learning and deutero-learning. Thus Bateson asserted in 1970 that mind must be understood to be "immanent in the . . . ecosystem . . . in the total evolutionary structure."[4]

General Systems Theory

Kenneth Boulding, author of the 1966 essay "The Economics of the Coming Spaceship Earth," is today seen as one of the pioneers of general systems theory, along with émigré Austrian biologist Ludwig von Bertalanffy. Boulding wrote to Bertalanffy:

> I seem to have come to much the same conclusion as you have reached, though approaching it from the direction of economics and the social sciences rather than from biology—that there is a body of what I have been calling "general empirical theory," or "general system theory" in your excellent terminology, which is of wide applicability in many different disciplines.

The phrase "general system theory" (later pluralized to "systems") was publicized by Bertalanffy in his book of the same title published in 1968, as an interdisciplinary field that aimed to describe and model the behavior of nested open systems wherever they were found—whether in cell, brain, society, electronic circuitry, or ecosystem. A general theory was feasible, Bertalanffy wrote, because there are "correspondences or isomorphisms in certain general aspects" of all open systems and their operation. These common features included nesting (the creation of a multi-leveled organizational structure of systems within systems), purposeful activity (homeostasis was the most commonly cited example of this), and a uniquely functional relationship between freedom and order. General systems theory, Bertalanffy maintained, offered an intellectually rigorous approach to what had been traditionally but somewhat mystically hailed as "wholeness" or "holism." It also implied a new scientific epistemology: "knowledge is not a simple approximation to 'truth' or 'reality,'" he wrote; "it is an interaction between knower and known, thus depending on a multiplicity of factors."[5]

Bertalanffy had attempted and failed to secure refugee status in the United States when the Anschluss occurred in 1938, so he had returned to Vienna, where he was persuaded to join the Nazi party. As a former Nazi collaborator, he later found it difficult to establish an intellectual foothold in postwar America.[6] But his reputation was redeemed by the enthusiastic support of Hungarian émigré Ervin Laszlo. As a Jewish child, Laszlo had survived the Nazi occupation of Budapest, and in 1947 had escaped with his mother from Soviet-occupied Hungary to Paris. He came to the United States in 1948 and began thinking and writing about systems theory in 1961.[7]

Laszlo's book *The Systems View of the World,* published in 1972 by George Braziller, encapsulated the insights of general systems theory for American readers.[8] Laszlo took as the starting point of his metaphysics the process philosophy of Alfred North Whitehead. In Whitehead's philosophy, the building blocks of nature are not particles but "events" defined by overlapping fluxes of energy in space-time. "Imagine a universe made up not of things in space and time, but of patterned flows extending throughout its reaches," Laszlo wrote.[9] Laszlo focused his attention on defining the dynamics that resulted in stable patterns amid that flow. These relationships constitute what he refers to as a "system."

Natural systems can be subatomic or macrocosmic, or of any size in between. On every level of scale, there is an organized communication among the system's constituent parts. That communication is what constitutes it as a system. For example, in an atom, which is among the least complex of natural systems, the communication is the forces of attraction and repulsion among subatomic particles. In human societies, which are much more complex, the mode of communication includes the feedback loops that we think of as culture.[10]

As communities of relationships rather than collections of objects, systems are not reducible to the properties of their constituent parts. Correspondingly, for the system to continue to exist, it is not necessary that its component parts be conserved. Electrons may escape an atom and new ones arrive; cells may die and be replaced; individual animals may also die and reproduce as the system renews itself. But "sets of relationships . . . are conserved."[11]

The behavior of every component of the system is shaped by the relationships of interdependency within the system. In order to be considered a system, wrote Laszlo, a group of entities must be connected by "effective, mutually qualifying interaction." This short phrase describes the function of feedback as a constraint. It includes three points: the interaction must have effects on the entities involved; those effects must be mutual, not unidirectional; and the mutual effect serves to constrain ("qualify") the actions of each component. "Systemicity," Laszlo asserted, "is imposed as a set of rules binding the parts among themselves."[12]

Because of the necessity of feedback dynamics, a system's components are never completely subsumed in a homogeneous mass. They must retain some individuality and "functional autonomy."[13] The characteristic form that systems take, as a result, is a lumpy coherence, like that of a body with organs, a cell with organelles, or a society with subgroups.

Cyberneticists and ecologists are almost exclusively interested in "open" systems. A system is open when, in addition to its internal communications, it is in communication with its surroundings. The dynamics structuring an open system enable it to maintain itself in a changing environment. The most common dynamic of such self-maintenance is homeostasis, which stabilizes the internal environment of the system despite fluctuating external conditions. Homeostasis is a "dynamic steady-state" produced by shifts in the behaviors of the system's various subsystems.[14] For example, human bodies maintain an internal temperature of around 98.6 degrees Fahrenheit; if they get hotter, the nervous system stimulates the eccrine glands to release sweat and promote cooling.

Another key point about open systems, according to Laszlo, is that they are almost always nested inside larger ones. This means that the "external" environment to which an open system responds is actually the dynamics structuring the next larger system in which it is a subsystem. Conversely, its "internal" environment is made up of smaller subsystems, each of which is responding to the dynamics created by its relations with the other subsystems as its own "external" environment. Every open system is therefore an interface, mediating between other open systems of lesser and greater size and complexity. The fundamental rule of general systems theory is therefore this: "If any given thing is to maintain itself in proper running condition, it must act as a subsystem within the total system which defines its energy supplies."[15]

The constraints imposed on a system by the larger system of which it is a part (the external environmental conditions, from a perspective within the smaller system) operate differently from those that govern the system internally. While the latter narrowly define the behaviors of the system's parts, the former are only macrodeterministic—that is, they are imposed on the system as a whole, not on its individual parts. This results in a condition of relative autonomy within the subsystem in responding to wholly new conditions. As Laszlo explained, "The parts have options; as long as . . . the requirements of systemic determination are met." Therefore, many complex open systems adapt to environmental conditions not only by adjusting their internal activities, but also by refining the means by which they adjust them. This is akin to Bateson's idea of deutero-learning. A shift in external conditions can lead to a reconfiguration of the constitutive dynamics within the system, as a new way to respond develops.[16] The shift does not just mobilize the system to restore homeostasis, but to improve its adaptability by achieving a better range of internal interactions. To extend the previous example, at some point the human body developed the mechanism of sweating to regulate its internal temperature; it is not a technique common to all animals.

This process of innovation is how general systems theory understands evolution. An initial indeterminacy as to how the system will respond to a macro-determining change in its environment is what makes its evolution possible. Freedom or indeterminacy is therefore crucial to the plasticity (that is, flexibility and versatility) that improves an open system's ability to sustain itself in the face of change. Like ecological thinkers associated with the environmentalist movement, Laszlo contrasted self-organizing systems to mechanistic systems in this respect. "Totally mechanistic systems have only two states: a functional one where all parts work in the rigorously determined manner, and a failing one where one or more parts have broken down," he wrote. "They lack the plasticity of natural systems, which act as dynamic, self-repairing wholes in regard to any deficiency."[17]

It is this dynamic, of indeterminacy resolving into purposeful action, that makes the functioning of self-organizing systems equivalent to mind. Laszlo asserted that "whether a system is physical or mental depends on the viewpoint of observation." From a perspective within the system, its actions are seen to be physical; from the outside, as the indeterminacy of how, precisely, it will respond to environmental stimuli is resolved by its assumption of a new state consistent with constraints, the process is one of decision making—hence, mental.[18] "Free systems," Paul Weiss wrote in 1973, are "problem solvers."[19]

As a survivor of Nazism and a refugee from Stalinism, Laszlo was not slow to apply this lesson of general systems theory in a critique of centralized political power. "Natural systems form something like democracies," he wrote, in which "there is a plan, but it is not a preestablished one. It sets forth the guidelines and lets . . . [indeterminacy] play the role of selector of alternative pathways for its realization. There is purpose without slavery, and freedom without anarchy." At the same time, the functional autonomy of the individual within such a social system should not be confused with "independence" in a libertarian sense. "A fully autonomous (independent) set of units would not constitute a system, only a heap," Laszlo admonished.[20] The laws of ecology implied that societies that were at once both democratic and aware of the environmental constraints within which they operated would be the strongest and most successful ones.

In this connection, Laszlo believed, environmentalism represented society's necessary acknowledgment of the macrodeterministic constraints placed on human behavior. As a central European introduced later in life to America's culture of consumer capitalism, he viewed it from a critical perspective. "The Western world tends to offer the values of affluence as the panacea for all social ills," he observed; but this behavior had resulted in an unsustainable level of resource consumption. Therefore, now "progress must be redefined, and that means a new system of values." Ecological thinking was as crucial to American strength and survival as was individual freedom. And given America's cultural history, Laszlo predicted, it would be more difficult to advance politically. "The supreme challenge of our age," he wrote with emphasis, "is to specify, *and learn to respect*, the objective norms of existence within the complex and delicately balanced hierarchic order that is both in us and around us."[21]

Meeting this challenge would require, he asserted, a new approach to knowledge. Our science was inadequate. The modern scientific method embraced a strictly limited form of empiricism. As a result, it was capable of exploring only the simple relationships that could be adequately tested and verified by experiment. The method had been developed in the Enlightenment era, in the context of a need to refute ideas about nature promulgated by the church and classical philosophy. But science today was hobbled by it. The time had come for science to leave behind its youthful insecurities and to evolve into a process of inquiry more open to possibility and more expansive in scope. In ethology, for instance, a holistic systems view must supplant the dominant behaviorism, making it possible for scientists to acknowledge and explore the subjectivity of animals.[22]

The identity of mental and physical processes, and the necessity of an intersubjective epistemology: these were the metaphysical lessons that Laszlo derived from general systems theory. The values that it implied were those of democracy and of an ecological ethic based on a "*reverence for natural systems*." Laszlo wrote that these principles provided the basis for a new humanism demanded by the times. The times were indeed desperate, but therefore also full of possibility, since in an open system, environmental pressures registering as constraints are the means to evolution.[23]

Coevolution

Ecological thinkers of the seventies understood "coevolution" as the most pervasive example of mind in nature. As opposed to orthodox Darwinian theory, which imagined evolution to result from the chance mutation of individuals that then propagated through a process of elimination of the unfit, the theory of coevolution conceptualized the process as a purposed (in a cybernetic sense) self-organization of the ecosystem. In his 1978 book *Thinking Animals*, Paul Shepard described it as the "evolution of the terrestrial mind." Adaptations were "decisions" of the ecosystem, made in real (what Norbert Wiener had called "Bergsonian") time.

The theory of coevolution emerged among biologists in the 1960s, as systems ecology began to overlap with population genetics.[24] Paul Ehrlich, who later wrote the environmental best seller *The Population Bomb*, in 1964 published one of the first scientific studies advancing the theory, suggesting the coevolution of certain butterflies and plants. In 1968 Arthur Koestler organized an Alpbach Symposium on the topic "Beyond Reductionism: New Perspectives in the Life Sciences," which brought together a number of distinguished thinkers in systems theory, including biologists Paul Weiss and Conrad Waddington; cognitive psychologists Jean Piaget and Jerome Bruner; and Ludwig von Bertalanffy.[25] At the symposium, coevolution emerged as a key concept demonstrating the immanent intelligence of self-organizing systems and its pervasiveness in nature.

Koestler's own contribution to the symposium summarized his idea of the "holon," first introduced in his 1967 book, *The Ghost in the Machine*. A holon is a unit useful in describing an open system nested in a larger system. As a system facing its environment, it is a whole; yet as a nested subsystem, it is a part. The external "environment" to which a holon responds is actually nothing other than its dynamic interactions with all the other holons that are each likewise a semi-autonomous subsystem of the one it is nested in.[26] Therefore, when a holon evolves, it never does so independently, but only in concert with those other subsystems.[27] This is the basic principle of coevolution.

The emerging science of epigenetics, which focused on the mysteries of protein-DNA interactions, offered evidence that evolution was governed by such feedback dynamics rather than by the simple mechanisms of chance and elimination embraced by orthodox Darwinists. The chances that a survival advantage would result from a random genetic mutation are almost infinitely small. Some other process must therefore be at work.[28] Epigenetics describes the mechanism by which feedback loops bring the chemical and behavioral patterns of an organism to bear, increasing the chances of the right mutation occurring in the right environmental context. Evolution, then, is best understood as an event not of random chance but of organized complexity.

Conrad Waddington, the originator of modern epigenetics theory, explained at Koestler's 1968 Alpbach Symposium how epigenetics supports the theory of coevolution. In studying evolution, it is important to make a distinction between phenotype and genotype. An ecological niche can be occupied by an organism with certain properties and capabilities; this is called its "phenotype." Genetic mutation and inheritance produce the organism's genetic code; this is called its "genotype." The ambiguity lies in the relationship of genotype to phenotype. Epigenetics revealed that proteins switch genes on and off, improving the fit of genotype to phenotype. As Waddington summarized, "The same genotype can [because of enzymes epigenetically activating and deactivating genes] . . . produce a number of phenotypes according to what the environment of the developing system has been."[29]

Reciprocally, "this means that if you start with a phenotype, as natural selection does, there is an essential indeterminacy in the relation between that phenotype and the genotype"—an indeterminacy that can only be resolved through non-genetic chemical processes initiated by the organism's responses to its environment, such as hormonal secretions or the chemical effects of its behaviors. In short, patterns of practice in a group of animals can shape their evolution. This is known as the "Baldwin effect," after turn-of-the-century biologist James Mark Baldwin. An organism's adaptation to its environment is first coded behaviorally and epigenetically, until such time as a genetic mutation hardwires it into the species' long-term memory (its genotype). Waddington concluded, "It remains true enough to say that . . . the genes in the organisms have been produced by random processes, but this is almost irrelevant" since "the randomness of the basic gene mutations is buried deep in the complexity" of environmental feedback dynamics.[30]

As a model of the cybernetic intelligence immanent in nature, coevolution implied that the competition readily observable in nature was but a subset of dynamics that registered as cooperation at other levels of complexity.[31] Paul Shepard explained that the interdependency of predator and prey species dictated that their brains and skills had to evolve more or less in tandem if both were to survive: "The predator must have an advantage but not reduce the prey too much." This systemic relationship resulted in "cycles of predator and prey whose dances become less and less random encounters and more and more choreographed" over the ages.[32]

The idea that mind was immanent in the organized complexity of nature, and discernible in such processes as coevolution, offered a holistic alternative to the Cartesian premise of a mind/body dualism, which ascribes intelligence only to a hypothetically transcendent human consciousness. Gregory Bateson in his 1967 lecture at the Dialectics of Liberation conference argued that human consciousness, although a useful adaptation, brought with it mental limitations as well as benefits. "On what principles does your mind select that which 'you' will be aware of?" Bateson asked. "I, the conscious I, see an unconsciously edited version." The conclusions arrived at by systemic intelligence, by contrast, were manifested in what physically happened as a consequence; this was not always synonymous with the conclusions reached by conscious thinking alone, which could overestimate the power of the will to control events. In an essay first published in 1971 titled "The Cybernetics of 'Self,'" Bateson asserted that "the total self-corrective unit which processes information, or, as I say, 'thinks' and 'acts' and 'decides,' is a *system* whose boundaries do not at all coincide with the boundaries . . . of what is popularly called the 'self' or 'consciousness.'"[33]

In the systemic view of intelligence, one's "mind" extends throughout the body and beyond it into one's physical surroundings. One implication of this is that the body participates in mental processes. "This network," Bateson wrote, "extends to include the pathways of all unconscious mentation—both autonomic and repressed, neural and hormonal." The thinking done by the unconscious body, he noted, was often referred to as "feelings," because it was typically "accompanied by visceral and other bodily sensations." Another, even more radical implication was that the "individual" mind, even as a subset of the larger ecosystem mind, was something of an arbitrary designation: more of a cultural convention than a tangible fact.[34]

Seeing evolution as a manifestation of systemic intelligence also highlights the significance of material form and physical pattern in mental processes, as the outcome of decisions or choices made by such an intelligence. Structure and pattern are both forms of information.[35] As the products of intelligent decision rather than of chance mutation, the shape and structure of an entity constitute a repository of its learning. Bertalanffy, speaking at Koestler's Alpbach Symposium, hypothesized that the "organizational laws" of coevolution explain why parallel forms exist among separately evolved animals. Analogous forms indicate underlying similarities in the dynamics of systemic interaction, even when found in taxonomically different parts of the system, or on different levels in nested systems ("self-similarity").[36]

This understanding of form was the conceptual basis for Godfrey Reggio's movie *Koyaanisqatsi*, which has been described as "one of the most frequently rented films on the American college circuit [and] an almost inevitable Earth Day presentation." The making of the film, with a soundtrack by Philip Glass, began in 1975, and it was released in 1983. Its title is an ancient Hopi word that Reggio translates as "life out of balance." Without actors or script, *Koyaanisqatsi* presents a visual study of the physical forms resulting from the dynamics of interaction structuring two intersecting ecologies: the natural world and industrial society. Reggio's film compares and contrasts the shapes and flows of the two systems, presenting them as "the larger patterns within which what we call individuality is subsumed." He uses slow-motion and time-lapse photography to collapse the differences in speed at which various processes occur, revealing their underlying similarities of form. In one well-known scene, for example, car traffic on urban freeways is filmed at night using an extreme long shot and time-lapse photography. The individual points of light created by headlights and taillights merge into flows that stop and start rhythmically, like blood flowing in veins and arteries.[37]

Gaia

Coevolution was offered as a guiding metaphor by Stewart Brand when he began promoting ecological thinking in the early seventies as the next phase of the sixties counterculture. Brand had first encountered the theory of coevolution at Stanford University in the late 1950s, where he took biology classes with Professor Paul Ehrlich. After reading Bateson's *Steps to an Ecology of Mind,* he was inspired to begin publishing the *CoEvolution Quarterly,* which he edited through 1984.[38]

In the summer of 1975, the *CoEvolution Quarterly* introduced its readership to the Gaia hypothesis of James Lovelock and Lynn Margulis, which imagined the entire biosphere, including its inorganic components, as a single intelligence or organism. In 1967 Lovelock and Dian Hitchcock had argued in Carl Sagan's journal *Icarus* that the Earth's atmosphere was "a dynamic extension of the biosphere," in that its levels of methane gas could only be accounted for by the presence of life processes. In 1972, however, Lovelock went further and suggested that such gas levels—along with multiple other inorganic factors in the environment such as atmospheric pressure, ocean salinity, and surface temperature—were not unintended by-products of life processes, but resulted from life's intelligent regulation of its inorganic environment through cybernetic control. "The biosphere interacts actively with the environment so as to hold it at an optimum of its choosing," he wrote. "Life at an early stage of its evolution acquired the capacity to control the global environment to suit its needs." He named this biospheric intelligence Gaia and described it as "a complex entity involving the Earth's biosphere, atmosphere, oceans, and soil; the totality constituting a feedback or cybernetic system which seeks an optimal physical and chemical environment for life on this planet."[39]

Having conceived of this radical idea, Lovelock in late 1970 began consulting with Lynn Margulis, a biologist at Boston University who had established a reputation for successfully supporting controversial coevolutionary hypotheses with scientific data. She became convinced of the truth of his hypothesis in 1972, and together they published three additional essays, all of which appeared in 1974.[40] These brought the Gaia hypothesis to the attention of the editor of the popular British science magazine *New Scientist,* which did a cover story, "The Quest for Gaia," in February 1975. Coverage in the *CoEvolution Quarterly* followed soon after.

It was in the *New Scientist* that Lovelock for the first time ventured to suggest in print that Gaia "seemed to exhibit the behaviour of a single organism, even a living creature." Although Margulis demurred from this characterization, Lovelock made the claim with even more conviction in a book-length exposition of the idea that he published in 1979, titled *Gaia: A New Look at Life on Earth.* In it, he described Gaia as "the largest living creature on Earth . . . a vast being who in her entirety has the power to maintain our planet as a fit and comfortable habitat for life." The idea quickly caught on in American popular culture. Fritjof Capra declared in his 1982 book *The Turning Point*: "The earth, then, is a living system; it functions not just *like* an organism but actually seems to *be* an organism—Gaia, a living planetary being."[41]

The image of Gaia inspired public intellectuals who embraced ecological thinking. Morris Berman wrote that if you "believe it is somehow all right to pollute Lake Erie until it loses its Mind, then you will go a little insane yourself, because you are a sub-Mind in a larger Mind that you have driven a bit crazy."[42] However, there were also passages in Lovelock's book that worked against the environmentalist message by suggesting that the problem of pollution had been overstated. Lovelock took an anti-alarmist viewpoint. The worst atmospheric pollution in Earth's history was the introduction of free oxygen, he wrote, which had made the evolution of life as we know it possible.

> When oxygen leaked into the air two aeons ago, the biosphere was like the crew of a stricken submarine needing all hands to rebuild the systems damaged or destroyed and at the same time threatened by an increasing concentration of poisonous gases in the air. Ingenuity triumphed and the danger was overcome, not in the human way by restoring the old order, but in the flexible Gaian way by adapting to change and converting a murderous intruder into a powerful friend.

Similarly, Lovelock believed, despite Barry Commoner's dire warnings in *Science and Survival*, that "a nuclear war of major proportions, although no less horrid for the participants and their allies, would not be the global devastation so often portrayed. Certainly it would not much disturb Gaia." Higher levels of radiation were not necessarily bad from an evolutionary standpoint.[43]

Lovelock elaborated on his ideas with flights of technocratic fantasy. In one scenario in *Gaia: A New Look at Life on Earth*, he imagined hydrogen bombs deployed to fend off an asteroid on a collision course with Earth. In another, scientists purposely released gases into the atmosphere in order to cool the planet. Such interventions were justified by the thought that humanity was by analogy the central nervous system of Gaia: the means by which she had finally become self-aware and added conscious action (via human intervention) to her arsenal of techniques for self-preservation.[44]

These aspects of Lovelock's writing exemplify a turn toward science fiction that some ecological thinkers took near the end of the seventies decade. Earth's finite resources, it seemed, might not, after all, mark the limits of human growth. Gaia might extend her reach to other planets. In 1984 Lovelock coauthored a full-blown science fiction novel, *The Greening of Mars*, in which humans make Mars habitable by polluting its atmosphere with chlorofluorocarbons. The novel's title was perhaps an ironic play on Charles Reich's classic account of the hippie counterculture, *The Greening of America*. The popular science fiction series by British author Douglas Adams, The Hitchhiker's Guide to the Galaxy, also featured familiar ecological ideas subsumed into a space-travel fantasy, in which the end of the world is just the beginning of the adventure. Stewart Brand's *CoEvolution Quarterly* was also a vehicle for this particular strain of ecological thinking in the latter part of the decade. In the fall of 1975, it promoted plans by Princeton physics professor Gerard O'Neill for a space colony that he anticipated would be inhabited by one million people by the year 2000. Among some readers such stories generated excitement about the possibility of creating artificially engineered "miniature Earth" environments that could be launched into space, effectively allowing Gaia to "seed" distant planets. Other readers, however, were dismayed.[45]

Such notions were diametrically opposed to the strategy of minimal intervention in the systemic control of natural processes that had been embraced by such thinkers as Rachel Carson and Gregory Bateson. Gary Snyder told an interviewer in 1977:

There are two kinds of earth consciousness: one is called global, the other we call planetary. The two are 180 degrees apart from each other, although on the surface they appear similar. "Global consciousness" is world-engineering-technocratic-utopian-centralization men in business suits who play world games in systems theory. . . . "Planetary thinking" is decentralist, seeks biological rather than technological solutions, and finds its teachers for its alternative possibilities as much in the transmitted skills of natural peoples of Papua and the headwaters of the Amazon as in the libraries of the high Occidental civilizations.

Adhering to "old-ways" ecological thinking and its focus on local ecosystems, Snyder found himself at odds with the new technological utopianism of "global consciousness," which, he feared, "ultimately would impose a not-so-benevolent technocracy on everything via a centralized system."[46]

Instead the idea of Gaia supported Snyder, as it had Bateson, in embracing a profound redefinition of the self in relation to nature. "We are interdependent energy fields," Snyder wrote. "Coyote the animal, Human Being the animal, Bear the animal, are just . . . shapes and functions assumed in the service of the evolution of the great biosphere-being, Gaia."[47] This insight, he felt, encouraged a psychological disposition that promoted environmentalism by moving individuals past a culturally reinforced fear of death. Death is not as fearsome when one "can accept his place in a process far greater than himself." In turn, if death is not feared, the desire to artificially ensure safety and stability decreases. That desire for artificial stability was what motivated unsustainable levels of consumption. Conversely, accepting the insecurity of a lifestyle dependent on the fluctuating cycles of nature made sustainable living thinkable. The American culture of competitive individualism ran contrary to this ethic by equating death with elimination rather than renewal. It was "counterproductive when the important insight for everyone is how to interact appropriately and understand the reciprocity of things, which is the actual model of life on earth."[48]

The potential social importance of poetry, in Snyder's view, lay in its ability to spread this ecological insight. For this purpose, he took Native American shamanism as a model for his poetics. Citing *Black Elk Speaks*, he credited the Plains Indians with the understanding that "trees, animals, mountains are in some sense individualized turbulence patterns, specific turbulence patterns of the energy flow that manifest themselves temporarily as discrete items, playing specific roles and then flowing back again." Through shamanistic practices like meditation, Snyder believed, the poet could intuit the subjectivities of other living systems and make audible the otherwise inarticulate voices of nature. In this way, poetry could complement the scientific approach of ecology, as a way of listening to nature. At a conference on the Rights of the Non-Human in the spring of 1974, Snyder told his audience:

> The Cahuilla Indians . . . said not everybody will do it, but almost anybody can, if he pays enough attention and is patient, hear a little voice from plants. . . . The shaman speaks for wild animals, the spirits of plants, the spirits of mountains, of watersheds. He or she sings for them. They sing through him.

Snyder lauded the ritual dances of the Pueblo Indians for providing them with a process by which their political leaders consulted with the non-human parts of the ecosystem by allowing "some individuals to step totally out of their human roles to put on the mask, costume, and mind of Bison, Bear, Squash, Corn, or Pleiades; to re-enter the human circle in that form and by song, mime, and dance, convey a greeting from the other realm." What was needed was the integration of such shamanic practices into modern American culture. The result, he ventured, would be akin to "a speech on the floor of Congress from a whale."[49]

THREE

Crying Indian

God Is Red. VINE DELORIA JR.

Gary Snyder believed that mainstream Americans had much to learn from Native Americans. "It takes a long time to get to know how to live in a region gently and easily and with a maximal annual *efficiency*," he explained to an interviewer in 1971, using the ecological definition of that term. "What we call the primitive is a mature [cultural] *system* with deep capacities for stability and protection built into it."[1]

The idea that Native Americans related to nature better than the European Americans who had learned how to exploit it most fully was widespread in 1970s popular culture. One manifestation of this was the emergence of the "crying Indian" as an iconic figure in films and advertising. In the crying Indian motif, environmentalism was linked to the importance of a kind of knowledge that came from outside mainstream American culture and its technologies. Crying was indicative of the Indian's special affective relationship with nature.

Because ecological thinking, like an earlier Romanticism, elevated feelings as a form of intelligence more complete than what was available to the "objective" intellect alone, the Indian's affective grasp of nature could be construed as truer than the perspective of Enlightenment science. Ecological thinking positioned intersubjectivity as an epistemological corrective to the scientific method and its rigorous rejection of subjective truths. In keeping with this idea, neo-paganism and ecofeminism valorized the traditional Native American worldview as one that located mind in nature. They equated this with an ecological wisdom that Western civilization had sacrificed in pursuit of a knowledge organized around the quest for power.

The fascination of many Americans in the seventies with forms of Native American spirituality led to an increase in the popular appropriation of traditional Native American practices. These performances were of varying degrees of legitimacy. But reciprocally, Native American intellectuals, including John Mohawk and Vine Deloria Jr., made use of the connection that had been forged between ecological thinking and Native American traditions in order to advance the cause of American Indian rights.

The Crying Indian

Most Americans who watched broadcast television in the early seventies remember the "crying Indian" public service announcement that premiered on the second Earth Day, in 1971. In the ad, actor Iron Eyes Cody, dressed as a Chippewa Indian, paddles a canoe through a pristine natural landscape that gradually gets more and more polluted. He drags his canoe up onto a litter-strewn shore and looks around him, as a voice-over pronounces, "Some people have a deep abiding respect for the natural beauty that *was once* this country." While the voice-over proceeds, the Indian is walking over to a busy freeway. "And some people *don't*," the voice-over continues, as a passenger in a speeding car flings a bag of trash that lands right on the Indian's feet. Cody looks up from the trash to stare pointedly into the camera, which zooms in to capture, in freeze-frame and extreme close-up, a lone tear escaping from his eye (see plate 1).

The image of the crying Indian simultaneously mobilizes two different ways of thinking about tears. One interpretive schema emphasizes their private and cathartic nature: they relieve the pressure of an interior pain.[2] From this perspective, the crying Indian means "*even* an Indian cries." The stoic icon of Western lore, legendary for his impassiveness in the face of physical pain, is nevertheless reduced to tears by the appalling levels of modern-day pollution. Although he is still stonily wordless, his involuntary tear betrays the extremity of his suffering.

Seen this way, the PSA simultaneously invokes and repurposes popular depictions of Native American men as fierce warriors, familiar from 1960s television and movie Westerns, from *Comanche Station* to *Mackenna's Gold*. Much of the PSA's cultural work in this respect is accomplished by the camera's attention to Cody's face, which reveals the inner struggle he will not otherwise express. But this message is also reinforced by the ad's music, which shifts midway through from a heroic martial theme marked by timpani, snare drums, and trumpet to a mournful theme in which string instruments predominate. The implication is that the mythically fierce Indian *has become* the crying Indian.

The other schema for making sense of tears constructs them as a social rather than as a private event, interpreting crying as a communicative behavior. Tears *signal* suffering and evoke empathy. When the Indian cries, we see ourselves in him, and our feelings reach out to him unconsciously. We commiserate with him and share affectively in his loss.[3]

Seen in this light, the iconic crying Indian is not stoic; to the contrary, he has an emotional connectedness ("*only* the Indian cries") that we aspire to emulate. A sequel to the original ad, featuring Iron Eyes Cody on horseback, emphasized this message by varying the voice-over to say, "The first American people loved the land; they held it in simple reverence. And in *some* Americans today that spirit is reborn." It is implied that the Native American, as an outsider, preserves a sensibility toward nature that the dominant culture has conditioned people away from, but that is in the process of being recovered.

The 1971 ad employs cinematographic techniques to emphasize this structural opposition between modern technology and Indian stewardship of the land. A double-exposure shot dramatizes the disparity between the Indian paddling his canoe and the industrial landscape into which he has wandered (fig. 4). Later, rack focus is used to shift our attention from a close-up of the Indian's face to the four lanes of traffic on the highway beside him, creating a jarring juxtaposition. When the callous car passenger throws litter directly onto the Indian's body, it literalizes a metaphor that the ad has been constructing all along: the Indian represents the Earth we are exploiting.

Multiple texts of the environmental movement emphasized the idea of an emotional connection between Native Americans and nature that mainstream Americans had lost. Stewart Udall wrote in the first chapter of his 1963 book *The Quiet Crisis*, titled "The Land Wisdom of the Indians," that "the conservation movement finds itself turning back to ancient Indian land ideas, to the Indian understanding that we are not outside of nature, but of it." Udall characterized the Native American

Figure 4. Double exposure in the Crying Indian PSA.

connection to nature as a fundamentally emotional one, based on a filial feeling of "reverence for the life-giving earth." He rhapsodized:

> The land was alive to his loving touch, and he, its son, was brother to all creatures. His feelings were made visible in medicine bundles and dance rhythms for rain. . . . [T]he Indian's emotional attachment for his woods, valleys, and prairies were the very essence of life.

Correspondingly, Udall interpreted modern America's litter and industrial pollution as symptomatic of an emotional pathology of the mainstream. "Our irreverent attitudes toward the land and our contempt for the Indians' stewardship concepts are nowhere more clearly revealed than in our penchant to pollute and litter and contaminate and blight once-attractive landscapes," he lamented.[4]

The Ecological Critique of the Scientific Method

The widespread belief that American Indians had an emotional connection to nature that mainstream Americans lacked emerged from a broader cultural critique of Western civilization and its science. The nineteenth-century Romantic rejection of Enlightenment attitudes was reinvigorated by the sixties counterculture.[5] In the seventies, ecological thinkers criticized the Enlightenment legacy of contemporary scientific discourse by condemning it as "Newtonian," "mechanistic," and "Cartesian."

By calling Enlightenment science "Newtonian" or "mechanistic," they referred to the fact that it posits a uniform physical world composed of discrete objects. Consistent with this assumption, the scientific method screens the truth of a proposition by subjecting it to a test of objective reproducibility. Unless one scientist's experience can be reproduced in a controlled laboratory setting by any other scientist, it cannot be received as scientific truth. The advantage of a mechanistic science is that, to the extent that it can be applied, natural processes become predictable. Any high school physics student should be able to tell you how fast any sphere will roll down any inclined plane.

Ecological thinkers rejected this epistemology as too limiting when confronted with the complex systems that constitute nature. Gregory Bateson wrote in 1971, "There is, among all scientists, a high value set upon prediction. . . . But prediction is a rather poor test of an hypothesis." This is because what the investigator encounters in nature is not an inanimate set of objects but an "intelligent" system, whose complex feedback dynamics are its most significant feature. The reproducibility of events under laboratory conditions is a sham because there is no way of completely restoring any past context and set of conditions of the system. Laboratory contrivances dramatizing a purported objectivity do so only by ignoring details that are deemed to be irrelevant, when in reality, "no man . . . has ever seen or experienced formless and unsorted matter; just as no man has ever seen or experienced a 'random' event." The most important truths might easily slip through the scientist's fingers through an insistence on such false reproducibility. "It is necessary, therefore, to look again among the fundamentals," Bateson wrote, "for an appropriate set of ideas against which we can test our heuristic hypotheses."[6]

The contrived setting of the laboratory situation could only falsify the results of experiments on intelligent entities. Not only did the scientific laboratory ignore the irreproducibility of systemic conditions and contexts, but it also ignored the context that it itself imposed. The pretense of creating a controlled environment eventuates in a uniquely unnatural context for the unfolding of natural processes. From a systems theory viewpoint, this context cannot help but affect the results of any experiment. "You can't really experiment with people, not in the lab you can't. It's doubtful you can do it with dogs," Bateson told Stewart Brand in 1973. "Because the experiment always puts a label on the context in which you are."[7] Intelligent entities respond to that context in addition to the intended stimulus. Consequently, when faced with an intelligence, laboratory science can only learn truths about its responses under laboratory conditions.

Ecological thinking's other main criticism of Enlightenment science was that it was Cartesian. In positing an inert, mechanistic physical universe, it was required to suppose a dualistic separation of mind and matter. Mind was imagined to be transcendent and ontologically different from physical matter. As such, it was necessarily restricted to conscious, rational thought. Emotions and reflexes, since they were bodily responses, could not be mental phenomena. In the Cartesian dualism, rational thinking was characterized by free will, whereas matter was subject to deterministic natural laws. Through scientific experiments, the rational mind of the investigator analyzed the qualities and behaviors of natural objects.[8]

By contrast, ecological thinking understood intelligence as a process involving material interactions. Visceral responses and "feelings" were indispensable aspects of it. Ervin Laszlo, writing in 1972, insisted that, contrary to the foundational premises of the scientific method, all interactions among living entities were ones in which feelings played a role. Hampered by its own method, science could not "use feeling to cognize." But Alfred North Whitehead's conception of selfhood as a "concrescence of prehensions" implied that feeling was the fundamental form of cognition. Whitehead, Laszlo recounted, accepted the bodily affects (including "anger, hatred, fear, terror, attraction, love, hunger, eagerness, and massive enjoyment") as expressions of an organismic intelligence.[9]

Morris Berman, who popularized Bateson's ideas in his 1981 book *The Reenchantment of the World,* emphasized that much of the knowledge apprehended through feelings was untranslatable to consciousness and its means of communication. The body and the unconscious were sources of a kind of knowledge that could not be put into words. It followed that sensual arts like music, oral poetry, and dance were the only paths of access to this knowledge. Berman wrote that they were therefore also the only sources of truly innovative ideas, since from an ecological viewpoint, innovation always proceeds from material causes. He predicted that feelings would play a much more central role in a future culture founded on ecological principles. "The inner psychic landscape of dreams, body language, art, dance, fantasy, and myth will play a large part in our attempt to understand and live in the world," he wrote. "These activities will now be seen as legitimate, and ultimately crucial, forms of knowledge."[10]

This is the discursive context in which the crying Indian trope emerged. In his book *The Turning Point,* Fritjof Capra explicitly linked the environmentalist image of the Indian to the critique of Cartesian epistemology. Because American Indians in their traditional cultures organized their lives around close interactions with natural systems, he wrote, they developed "a highly refined awareness of the environment"; and this familiarity bred an intuitive understanding of its complex processes, even if their cultures did not include what we would call a "scientific" or causal understanding of them. Ultimately the complexity of nature was beyond scientific understanding in any case, Capra asserted, since science relied exclusively on conscious apprehensions and reproducible effects. Western culture would therefore be wise to follow the Native Americans' lead and make better use of mental resources, like premonitions and intuitions, that the Scientific Revolution had rejected. "Ecological awareness arises from an *intuition* of nonlinear systems," he explained; and "such intuitive wisdom is characteristic of traditional, nonliterate cultures, especially of American Indian cultures."[11]

Neo-Paganism and Ecofeminism

Gary Snyder wrote in a 1975 essay titled "The Politics of Ethnopoetics" that poets promoting ecological thinking "were always 'pagans.' . . . The devil is, after all, not the devil at all, he is the miming elk shaman dancer at Trois Frères, with elk antlers and a pelt on his back, and what he's doing has to do with animal fertility in the springtime."[12] As Snyder's writings indicate, in ecological circles the critique of science merged with a critical reassessment of the Judeo-Christian religious tradition. According to this line of thinking, a religious precedent existed to the reductive Newtonian and Cartesian view of nature: medieval Christianity's dualistic distinction between the body and the soul. Like Cartesianism, this religious dualism discouraged the recognition of intelligence in nature. It derived, its critics hypothesized, from a fear of death that prompted desire for an everlasting life that was transcendent of matter. But its effect was to denigrate the physical and sexual aspects of life, in favor of a false purity of mind and soul. It denied legitimacy to the animism of ancient and Native American religions and vilified their gods.[13]

Many Americans in the seventies embraced an alternative religiosity based on the idea of a continuum between body and spirit, which they located in Native American and pagan traditions. In a study titled *Drawing Down the Moon*, ethnographer Margot Adler chronicled the rise of neo-paganism in America during the decade. Her interviewees typically expressed a "reverence for nature" and an "animistic" belief that everything "from rocks and trees to dreams" was a manifestation of the same living force. Ecofeminist writer Elizabeth Dodson Gray similarly called for a spirituality consistent with ecological thinking and rooted in the understanding that "all life, both physical and psychic, is unified in what we have been calling 'the dance of energy.' . . . God too is a biospiritual unity."[14]

The environmentalist aspect of this critique was most influentially articulated by medieval historian Lynn White, in an essay first published in 1967. White's "The Historical Roots of Our Ecologic Crisis" argued that Christianity's rout of paganism in medieval times laid the metaphysical foundation for the later Scientific and Industrial Revolutions. "The victory of Christianity over paganism was the greatest psychic revolution in the history of our culture," he asserted. "By destroying pagan animism, Christianity made it possible to exploit nature in a mood of *indifference to the feelings* of natural objects." Like the later crying Indian PSA, what White emphasized was Western civilization's lack of empathy with the Earth and other living beings. Moreover, he wrote, the "disastrous ecological backlash" that had resulted from this attitude could not be adequately addressed by a science that was itself created in the image of that medieval Christian dualism. Saint Francis of Assisi alone had pointed the right way, when he championed the existence of souls in animals.[15]

Originally published in the journal *Science*, White's essay was claimed by the environmental movement and repeatedly anthologized. David Brower's activist organization Friends of the Earth included it in *The Environmental Handbook* that they prepared for distribution on the first Earth Day in 1970. Ian Barbour included it in his 1973 textbook *Western Man and Environmental Ethics*. Ecofeminist Elizabeth Dodson Gray's book-length critique of science and Christianity as the twin tools of patriarchy (first published in 1978 then reissued in 1979 under a new title, *Green Paradise Lost*) was an expanded version of an essay that she had first written in 1974 called "The Psycho-Sexual Roots of Our Ecological Crisis"—a title echoing White's.[16]

Ecofeminism perceived in men's attempted dominion over nature the same impulse behind their attempted domination of women. Both relationships were characterized, Gray asserted, by a desire for absolute control instead of the shared control characteristic of feedback interactions.[17] Gray originally titled her book *Why the Green Nigger?*—a purposely provocative title that drew a parallel between the patriarchal subjugation of women and nature, and the slaveholder's denigration of the African slave. She understood all three forms of domination to be manifestations of a dualistic separation of spirit and matter, expressed in a sustained structure of cultural binaries: master/slave; man/woman; mind/body; independence/dependency; and science/nature.

The logic of patriarchy, Gray wrote, ascribed to the ideal man all of the positive poles of these binaries. The culturally valorized version of masculinity therefore connoted the power to rule one's body through the will; to rationally control one's emotions; and to renounce interdependency in pursuit of independence. This ideal man was, as a result, a self-alienated individual—a "tyrant" over both himself and others. His pathological selfhood was nevertheless assumed to be the norm of human subjectivity in scientific, political, philosophical, theological, and legal discourses.[18]

As masculinity's antithesis, femininity was denigrated as weak, emotional, carnal, and dependent. The feminine self was discursively constructed as unworthy of full personhood. A "rational" and "objective" science subjected nature to the same treatment. The Earth and its plants and animals were objectified and exploited like the body of a raped woman—that is, without empathy or acknowledgment of a shared personhood.[19]

Gray found an alternative to this patriarchal ideology in ecology's vision of systemic interconnectedness. Just as all learning and the knowledge it produces are necessarily interactive, she wrote, so, too, there is no such thing as transcendence or independence. She concluded, "When we honor *our bodies and the body of the earth*, we shall honor women and all life in a new way."[20] This was the basic truth of ecofeminism, connecting its environmentalist and feminist analyses.

Ecofeminism's elevation of the discursively "feminine" participated in ecology's epistemological rehabilitation of feelings. Gray explained that as part of repairing our severed connections to our physical bodies and the natural world, we must relearn to value feelings as a bodily form of intelligence:

> As a part of this integration process, we must struggle also to glue together again our feeling and our thinking. Feeling has been so downgraded in our culture (as 'feminine') and thinking (since it was declared masculine) so elevated into the

clouds of abstraction, that it is difficult to recall that humans have a thought every time we have a feeling and a feeling with every thought. Thinking and feeling are contiguous human processes constantly interpenetrating one another as we experience our lives.

Whereas Enlightenment discourses devalued a woman's wisdom because her female body was closely associated with natural processes like menstruation, pregnancy, and lactation, from an ecofeminist perspective this same involvement could be understood as bringing her a heightened awareness or intelligence. In this respect, Gray asserted, men were limited biologically as well as by social conditioning. Feminism could help to restore ecological wisdom by reasserting the female perspective in the halls of power.[21]

What the female perspective taught, Gray believed, were the values of diversity and interdependence. "What ecology helps us see is an earth covered with a vast array of ecosystems, both large and small, continuously interacting with one another in many ways which are not immediately obvious to humans," she wrote. Everything was potentially of value in ways that humans should not presume that they understood. Like nature, society was also a system organized by complex feedback loops, which depended for their operation on diversity rather than homogeneity. Therefore, instead of hewing to a social Darwinism that imagined progress as the elimination of the superfluous, social policy should be dictated by a "whole-system ethic" acknowledging that "our interdependence is so complex that . . . it is foolish to destroy *any* part of what we depend upon but do not fully understand."[22]

Neo-paganism and ecofeminism used ecological thinking to reinforce their rejection of culturally hegemonic dualisms in favor of seeing mind and spirit as immanent in nature. These cultural movements therefore contributed to the growing popular interest in traditional Native American belief systems that were seen to emphasize humanity's oneness with nature. Gray wrote that American Indians had more ecological wisdom than mainstream Americans; and she quoted a Sioux proverb as exemplary of the worldview she advocated. Both she and Gary Snyder took the Gaia hypothesis as science's belated rediscovery of what archaic religions had long revered as the Earth Goddess or Earth Mother. Snyder was convinced that the time was right for a definitive cultural transition toward traditional Native American values. He wrote in 1972, "Here is a generation of white people finally ready to learn from the Elders. How to live [as] . . . Natives of Turtle Island."[23]

Inheriting Native Ways

Philip Deloria has chronicled and critiqued the history of such fantasies in his book *Playing Indian*. From the Boston Tea Party onward, Deloria observes, European Americans have used imaginary Indians to hypothesize what it would mean to live differently from the way they do. But seldom have those who are playing Indian interacted with living Native Americans, since typically they have found the reality inconveniently divergent from the ideal they choose to imagine. Instead, they base their play on the premise that the real Indians have all but vanished.[24] In this sense, the crying Indian of the 1970s can be seen as a version of the vanishing Indian. In the same way that Americans of the late nineteenth and early twentieth centuries penned and painted elegies to the "vanishing" Indian cultures that they believed their "winning of the West" had left on the verge of extinction, so Americans of the seventies paid homage to the Native American as an image of the nature they feared they had all but destroyed.[25]

The mainstream appropriation of Native American ways took various forms. One mostly frowned upon by Native Americans themselves was the wholesale adoption of Native American identities by people who were not of Native American heritage. Among the whites who played Indian in this fashion was the actor Iron Eyes Cody, star of the crying Indian ads. He was born to Italian American parents in southern Louisiana. But until his death he insisted that he was born in Oklahoma Territory to a Cherokee father and a Cree mother.[26]

Two more of the most well-known "Native American" shamans of the era who were later exposed as imaginary Indians were Don Juan Matus and Jamake Mamake Highwater. Psychedelic guru Carlos Castaneda published several books beginning in 1968 that purportedly recounted his tutelage in traditional wisdom and drug use by a Yaqui elder named Don Juan Matus. The first book in the series was titled *The Teachings of Don Juan: A Yaqui Way of Knowledge*. When seekers of psychedelic enlightenment made it a best seller, it was followed by *A Separate Reality: Further Conversations with Don Juan* (1971), *Journey to Ixtlan: The Lessons of Don Juan* (1972); *Tales of Power* (1974), *The Second Ring of Power* (1977), *The Eagle's Gift* (1981), and still others. Castaneda insisted that his books were the result of anthropological fieldwork in Arizona and Mexico, and on that basis he was awarded a master's degree and then a doctorate in anthropology from UCLA. The popularity of his books made him a millionaire and a celebrity. However, after a 1973 cover story in *Time* magazine strongly suggested that his works were fictional, Castaneda retired from public life, announced that Don Juan had died, and refused further interviews. A more thorough debunking by Richard de Mille in 1980 traced many of Don Juan's utterances to books that Castaneda had access to through the University of California library system.[27]

Jamake Mamake Highwater was another Native American persona that its creator invented in order to capitalize on the cultural trend. Jackie Marks was born to Jewish parents in Los Angeles, but he passed as Native American beginning in 1969. Claiming Blackfoot and Cree heritage, he published several books that purported to offer an insider's perspective on Native American culture and spirituality. Among his most celebrated works were the children's novel *Anpao: An American Indian Odyssey*, which received a Newbery Honor in 1977; and *The Primal Mind: Vision and Reality in Indian America*, which was published in 1981. *The Primal Mind* attempted to convey the Indian's sense of oneness with nature through an eclectic mix of art history, literary and music criticism, and anthropology (much of it apparently gleaned from essays in Dennis and Barbara Tedlock's 1975 book, *Teachings from the American Earth: Indian Religion and Philosophy*). The Corporation for Public Broadcasting was preparing to produce a documentary based on *The Primal Mind* when articles in the *Washington Post* and *Akwesasne Notes* (the influential newspaper of the Mohawk Nation) exposed Highwater in 1984 as a white man playing Indian.[28]

Historian Alice Kehoe refers to such impostors as "plastic medicine men," invoking plastic as a material to emphasize their inauthenticity. But the phrase is also apt in relation to the other meaning of plastic, that of a comparatively permanent transformation in response to surrounding pressures. With this expanded definition, it becomes possible to identify other plastic medicine men in seventies popular culture who did not falsely inhabit American Indian identities but who nevertheless imagined themselves as "inheriting" an Indian cultural legacy, like adopted children.[29]

In 1978 Tom Brown published the memoir *The Tracker*, in which he recounted his extraordinary upbringing in the Pine Barrens of New Jersey in the late 1950s and early 1960s. While his parents turned a blind eye, he spent days and even weeks in the wild, learning tracking and survival skills under the tutelage of his friend Rick's Apache grandfather. The grandfather, Stalking Wolf, was, according to Brown, among the last generation of boys raised in the traditional Apache way before their conquest by the U.S. Army. He put Rick and Tom through the training that he had received in his youth, teaching them to stalk deer and bear and to survive blizzards and other natural hazards. After Stalking Wolf died in 1967, Rick went to Europe and died in an accident; but Tom survived to claim Stalking Wolf's legacy and become a vigilante avenger against deer poachers and other violators of nature's sanctity.[30]

In the essay "Ancestors," included in Michael Tobias's collection *Deep Ecology,* Brown described searching for Stalking Wolf in Arizona around the time that the latter died. He tells Stalking Wolf's Navajo neighbors, "He is my Grandfather." After a surreal desert journey marked by dreams or hallucinations, a vision leads him to Stalking Wolf's medicine bundle. This cements his conviction that he had been chosen to inherit his legacy: "I had become Stalking Wolf. He'd given me everything. I felt inadequate, but grateful."[31]

The dynamics structuring Tom Brown's vanishing Indian tale are similar to those that define the spiritual journey of Tyler, the fictional protagonist of the 1983 Disney film *Never Cry Wolf,* directed by Carroll Ballard. Tyler's character was based on Canadian environmentalist Farley Mowat, from whose memoirs and stories the screenplay was derived. In the Disney film, Tyler is an ecologist sent to the Arctic by the government to justify a proposed wolf-extermination program by scientifically documenting that wolves are decimating the caribou herd. Instead, echoing Aldo Leopold's famous essay "Thinking Like a Mountain," he realizes that the fault lies not with the wolves but with human incursion. The path to this recognition is coincident with his own spiritual journey. "Abandoned" by mainstream civilization, he is "adopted" by an Inuit named Ootek, who appears and disappears mysteriously and who teaches him traditional wisdom. Meanwhile Ootek's own blood relative—in a chiastic structure—"falls" away into the values of mainstream civilization.

Tyler's spiritual growth is symbolized in part by his changed attitude toward Rosie, a cowboy-style frontiersman who at the movie's outset appears to epitomize manly adventurousness and resourcefulness, but by the end of the movie is revealed to be a crass exploiter of nature. By contrast, Ootek understands what the wolves are saying to each other when they howl. Ootek initiates Tyler into the Native American understanding of the wolf as a spirit animal sent by the Earth Mother to maintain the health of the caribou herd. Meanwhile Ootek's nephew Mike follows in Rosie's footsteps: he kills the pair of Arctic wolves that

Tyler has befriended in order to sell their pelts and justifies his act by invoking the "survival of the fittest." In the end, Ootek and Tyler accept their respective losses and journey on together.

To some scholars it will seem most important to observe that Tyler and the Tracker exercised white privilege when they laid claim to voicing the American Indian perspective. But I think it is also of value to ask what they thought they had learned from their Native American mentors. These accounts ground in specific details many of the abstract concepts of ecological thinkers who imagined Native American cultures as models for the future. The abstractions of ecological thinking are evident, as when Brown wrote, "We learned a world view in which Nature is a being, larger than the sum of all creatures, and can be seen best in the flow of its interactions."[32] But they also offer more concrete information as to how this was imagined to work close-up, and how the Native American relationship to nature differed from that of America's cultural mainstream as a consequence.

What emerges in both narratives is an alternative to the scientific method that Andrew Pickering in *The Cybernetic Brain* named a "performative epistemology." Pickering associates this epistemology with a "nonmodern" (so as not to specify either premodern or postmodern) outlook, commensurate with a "vision of the world as a place of continuing interlinked performances." Among the concrete principles of this epistemology that emerge in both *Never Cry Wolf* and *The Tracker* are that (1) the knowledge gained is always contingent on circumstances; because (2) it is rooted in intersubjective interactions such as emulation and empathy; and that (3) it demands a quality of attention that is different from the modern norm, but which other animals typically use to relate to their surroundings. Brown described this quality of attention in *The Tracker* as "a constant refocusing between minute detail and the . . . whole pattern." By adopting this mode of awareness, Tyler and the Tracker become (as they claim) attuned to other animals, and also to the natural processes to which those animals are attuned.[33]

In the film *Never Cry Wolf*, Tyler rejects his assigned role as objective scientific observer and instead adopts the epistemological strategies of emulation and empathy—what is colloquially called walking a mile in another's shoes. Tyler attempts to become like the wolves he wants to know about, whom he describes as "traveling through their environment alert and attuned." He begins emulating them by marking his territory with urine; then he mimics their howls on his oboe; and he eats only what they eat, even when that turns out to be mice. At the film's climax, his performative epistemology reaches the point where he is running naked across the tundra with the wolf pack, to join them in bringing down a weak caribou. In the book version, Mowat also described "learn[ing] to nap like a wolf"; the first time he tried it, he fell asleep for several hours, but "the fault was mine," he explained, "for I had failed to imitate *all* the actions of a dozing wolf, and, as I eventually discovered, the business of curling up to start with, and spinning about after each nap, was vital to success. I don't know why this is so."[34]

Such emulation opens Tyler to an experience of empathy with the wolves. His participation in their lives leads him to see surprising similarities between theirs and his own. He develops an emotional bond with the wolves and an appreciation for their individual personalities. They become persons or subjects in his moral universe. The cultural work that director Carroll Ballard takes on in the Disney film is that of bringing the viewer to share Tyler's perspective in this regard. The movie works back and forth between Tyler's discoveries and his culturally reinforced fear of wolves, which the viewers are also positioned to share. In the film's climactic scene, Tyler is sleeping naked in the grass as the camera creeps up on him; the editing cuts between these shots of him and shots of wolves hiding in the grass. The viewer's assumption is that the wolves are now hunting Tyler. And when Tyler wakes to find himself among a herd of caribou, the viewer at first takes this as a justification of his traditional fears. By the scene's conclusion, however, Tyler's place in the drama of the hunt has been reconfigured. He has identified with the wolves, and they with him. As a result, the wolves accept his aid in the hunt, and he takes his share of the kill.

In *The Tracker*, Stalking Wolf also understands knowledge to be available only through the dynamics of interaction. Its content is inextricable from each person's process of acquiring it. Therefore Stalking Wolf's pedagogy consists entirely of guidance on the level of what Gregory Bateson called deutero-learning. "He taught us how to learn," Brown wrote; he "led us to situations where we could learn certain principles, but once he had shown them to us, it was up to us."[35]

What Brown primarily learns under Stalking Wolf's tutelage is a quality of attention. Early in their training as trackers, Tom and Rick are too purposeful in their attentiveness; but over time they learn how "never so [to] focus our attention that we were not also aware of the larger pattern around us." Brown contrasted this to the quality of attention demanded by his teachers at school:

> Where our schools were forcing us to pay total attention, Stalking Wolf was teaching us intermittent attention, a constant refocusing between minute detail and the . . .

whole pattern of the woods. . . . The more we learned to *let our attention wander and come to rest on the thing at hand just often enough to catch the disturbances* the better we became as trackers and as observers of the woods.[36]

This quality of attention is characteristic of many cultural practices associated with ecological thinking. It reappears in a variety of forms in the seventies culture of feedback. Experimental composers, from Pauline Oliveros and Terry Riley to Brian Eno, aspired to cultivate it by musical means. Biofeedback therapy conditioned it through the production of alpha waves. Gary Snyder in 1973 described it as the basis of his poetic method, which aimed at intuiting the subjectivity of other intelligent systems. "You have to practice a kind of detached and careful but really relaxed inattention," he told an interviewer, so that "the conscious mind temporarily relinquishes its self-importance."[37]

Brown claimed that this quality of attention, cultivated through years of practice in the wild, enabled Stalking Wolf to see and hear the animals of the Pine Barrens as they went about their lives. By this means, the animals "told" Stalking Wolf what was going on in the woods—not by speaking with him like Doctor Dolittle, but through their actions and the traces that their actions left behind. As Brown recounted:

Once while we were walking in the woods, . . . [w]e went under a tree while Rick was looking up and Stalking Wolf said, "Let it sleep." He kept walking but we begged him to tell us *what* was sleeping. He nodded at a Great Horned Owl on a branch above our heads. I was sure he had not looked up as we had approached the tree, or as we passed under it, . . . [so] [w]e pleaded with him to tell us how he knew, and he said that the mice had told him. When we wanted to know *what* the mice had told him, he said to go and ask the mice.

So Rick and I spent the next couple of weeks studying mice. Thereafter, we watched them to see what they were doing at that time of year, and we found that Stalking Wolf was right. Once we knew the pattern of life for the mice, we knew the pattern of life for everything that eats mice, and we could generalize to most other small rodents until we had a chance to study them the same way. We also knew to some extent the lives of the things-that-ate-the-things-that-ate-the-mice.[38]

Brown referred to this moment of insight as an instance of what Stalking Wolf called "good medicine." Good medicine and bad medicine, in his view, were feedback communications from an intelligence akin to James Lovelock's Gaia, that Brown called "the spirit-that-moves-through-all-things . . . some more complex pattern, something within my perception but beyond my complete understanding." Good medicine came when nature rewarded you for your right choices by revealing itself. As Brown explained, "The mice were good medicine. They led us beyond their mystery to the mystery of the way the lives of the animals were interdependent. They led us to an idea of how the whole fabric meshes together." Bad medicine, like being trapped for days in a tree by a pack of wild dogs, happened as a corrective when he and Rick were "out of [their] proper sphere of actions." Such negative feedback was "a sign of disfavor for the spirit-that-moves-through-all-things."[39]

Native American Intellectuals and Ecological Thinking

It is difficult to say to what extent Tom Brown's role as a white man playing Indian distorted his representation of Stalking Wolf's relationship to nature. Farley Mowat is a similarly problematic figure as a writer of nonfiction. He preferred to distill, conflate, and even invent details in order to get his point across. Particularly in his earlier works, he was prone to comic exaggeration and ironic understatement. In the seventies, a number of his earlier works were republished in revised editions, after outcries that he had distorted the facts. Yet he obstinately defended his stories even in those revised editions; in the 1973 reissue of *Never Cry Wolf,* he asserted, "It is my practice never to allow facts to interfere with truth."[40] He justified the liberties that he took in his storytelling on the basis of his greater obligation to convey the plight of the inland Inuit to the world. He had learned enough of their language and their ways to be able to see the world through their eyes, he wrote. "The People lent me their eyes so that I might see what white men have tried not to see. Now I, in turn, have lent the People my voice so that white men shall hear the words the Ihalmiut cannot speak for themselves."[41]

What is easier to ascertain is that Native American intellectuals of the seventies embraced ecological thinking as a means of expressing the cultural values that they were fighting to preserve. They were willing to meet the best of the "plastic medicine men" on this middle ground.[42]

Under the editorship of John Mohawk, *Akwesasne Notes*, the official newspaper of the Mohawk Nation in upstate New York, boasted over 70,000 subscribers from all over the country. Its editorial position, dovetailing with the environmental critique of Western technology, was that political and environmental crises worldwide had been caused by European domination and the lifestyle it had spread. "Western technology and the people who have employed it have been the most amazingly destructive forces in all of human history," Mohawk wrote.[43]

The adoption of traditional Native American attitudes toward nature, Mohawk believed, offered the only way out of the current crisis. In 1977 *Akwesasne Notes* sent delegates to a United Nations conference in Geneva, Switzerland, to deliver an address titled "Spiritualism: The Highest Form of Political Consciousness." It was published by Mohawk in 1978 as part of a book-length manifesto authored by "Iroquois Indians" with the title *A Basic Call to Consciousness*. The delegates stated:

> The traditional Native peoples hold the key to the reversal of the processes in Western Civilization which hold the promise of unimaginable future suffering and destruction. Spiritualism is the highest form of political consciousness. And we, the native peoples of the Western Hemisphere, are among the world's surviving proprietors of that kind of consciousness.[44]

They went on to describe this consciousness using language reminiscent of Ervin Laszlo's general systems theory: "All living things are spiritual beings. Spirits can be expressed as energy forms manifested in matter. A blade of grass is an energy form manifested in matter—grass matter. The spirit of the grass is that unseen force which produces the species of grass." Writing in *Akwesasne Notes* in 1973, Gayle High Pine contrasted this Native American perspective to the Christian religious notion of transcendent spirit:

> Unlike Christians, who dichotomize the spiritual and the physical, put religion in its compartment, and call the physical world evil and a mere preparation for a world to come, we recognize the 'spiritual' and the 'physical' as one—without Westerners' dichotomies between God and humankind, God and nature, nature and humankind, we are close and intimate and warm with Mother Earth and the Great Spirit.

High Pine's thinking followed the same lines as that of environmentalists, neo-pagans, and ecofeminists who demanded a rejection of the Christian and Cartesian dualisms and a return to Native Americans' emotional connection with nature.[45]

"Spiritualism: The Highest Form of Political Consciousness" also included a cogent historical analysis of the destructive path of Western civilization as a feedback loop between technology and power, characterizing it as "a death path for which their own culture has no viable answers. When faced with the reality of their own destructiveness, they can only go forward into areas of more efficient destruction." Beginning further back in history than Lynn White, the delegates to Geneva identified the historical roots of the ecological crisis not with medieval Christianity, but with the domestication of animals. This was Western civilization's first selection of purposeful control and predictable plenty at the expense of natural constraints. From a Native American perspective, it was an unethical usurpation of power. "Until herding, humans depended on nature for the reproductive powers of the animal world. With the advent of herding, humans assumed the functions which had for all time been the functions of the spirits of the animals." The advent of this exploitative ethos led to the development of agricultural technologies, which led in turn to improved tools and weaponry, which then led to patriarchy and centralized, hierarchical social authorities; and by extension, to imperialism and colonialism. Then, beginning in Roman times, "Christianity . . . effectively de-spiritualized the European world." The Christian dualism was developed through Enlightenment science to an almost complete conquest of nature during the Industrial Revolution, resulting in extensive pollution and mass extinctions. Now, the delegates concluded, "the air is foul, the waters poisoned, the trees dying, the animals are disappearing. We think even the systems of weather are changing. Our ancient teaching warned us that if Man interfered with the Natural Laws, these things would come to be."[46]

As political advocacy, the delegates' address effectively hitched the cause of Native American rights to the rising star of environmental politics. "Our essential message to the world is a basic call to consciousness. The destruction of the Native cultures and people is the same process which has destroyed and is destroying life on this planet," they exhorted. Correspondingly, a widespread adoption of the traditional Native American worldview would restore basic rights and wholeness to all creatures—human beings, animals, and plants.

> The people who are living on this planet need to break with the narrow concept of human liberation, and begin to see liberation as something which needs to be extended to the whole of the Natural World. What is needed is the liberation of all the things that support Life—the air, the waters, the trees—all the things which support the sacred web of Life.[47]

Lakota activist and intellectual Vine Deloria Jr. also adopted ecological thinking as a means of Native American advocacy. In his 1973 book, *God Is Red: A Native View of Religion*, he wrote that Native American spirituality offered the necessary alternative to the Christian credo of humanity's dominion over nature. "In the area of ecology more than any other field," he asserted, "the Christian churches appear to be helpless."[48]

In a follow-up work published in 1979 titled *The Metaphysics of Modern Existence*, Deloria further developed the connections between a traditional Native American worldview and general systems theory. There he linked his earlier critique of Judeo-Christianity's transcendent God to the ecological thinking of Gregory Bateson, quoting Bateson's *Steps to an Ecology of Mind*:

> If you put God outside and set him vis-à-vis his creation and if you have the idea that you are created in his image, you will logically and naturally see yourself as outside and against the things around you. And as you arrogate all mind to yourself, you will see the world around you as mindless and therefore not entitled to moral or ethical consideration.[49]

Deloria characterized the American Indian worldview as, by contrast, coevolutionary: "each species gaining an identity and meaning as it forms a part of the complex whole."[50]

Deloria argued that environmentalism's belated rediscovery of the usefulness of traditional American Indian practices such as controlled burn suggested that Native American cultures were highly advanced rather than "primitive." Western civilization had made a devil's bargain by amassing power in return for becoming alienated from nature.

> Primitive peoples . . . do not deserve the scorn that Western thinkers, philosophers and theologians, especially, have heaped upon them. If they did not construct massive scientific technologies and immense bureaucracies, neither did they create the isolated individual helplessly gripped by supraindividual forces.

Following other ecological thinkers, Deloria drew connections between the Christian dualism and the limitations of Cartesian or Newtonian science. Due to a metaphysics plagued by this dualism, Enlightenment science had developed "by its very nature incomplete and narrow, avoiding a confrontation with the complexity of phenomena." However, modern physics, as interpreted through the process philosophy of Alfred North Whitehead, opened the door to another metaphysics in which all knowledge was understood to result from interactions and relationships. "Western science has thus arrived at precisely the starting point of non-Western peoples," Deloria scoffed.[51]

Deloria elaborated on plant sentience as one example of the knowledge that Native Americans grasped intuitively, that Enlightenment science found impossible to accept, and that ecological thinking would validate. He wrote in *The Metaphysics of Modern Existence*:

> Some scientists have been able to show electrical impulses that would indicate the presence of joy and pain within plants. . . . The current popular fad of playing music to assist plants in achieving better growth stems from this initial perception of the psychic nature of plants.

In this connection, he cited a study by biologist Frank Brown, who in 1959 established the sensitivity of plants to "still unknown subtle and highly pervasive forces" differentiating times and possibly places. Potato plants sealed hermetically under ostensibly constant laboratory conditions nevertheless varied their rates of metabolism in ways corresponding to each day's completely unpredictable levels of background radiation, outdoor air temperature, and barometric pressure. There was no way of accounting scientifically for the plants' responsiveness to stimuli from which they were supposedly completely insulated. Brown concluded that they were responding to subtle signals of which scientists were unaware, and which invalidated the fundamental epistemological premise of experimental physiology: that of reproducibility under "constant conditions."[52]

Optimistically referencing such studies, the psychedelic youth counterculture, and the works of Gary Snyder and Gregory Bateson, Deloria predicted that "the newly emerging view of the world will support and illuminate Indian traditions." As an Indian rights activist, he dared to hope that once the systems view of the world was generally adopted, Native Americans would no longer suffer social disadvantage for adhering to their traditional versions of it. "Economic and political decisions will begin to reflect a more comprehensive and intelligent view of the world and of our own species," he predicted, "thereby taking the pressure, in a political and economic sense, away from the surviving primitive and tribal peoples."[53] It was not yet too late to make that change.

FOUR

Talking with Plants

To know a place well means, first and foremost, I think, to know plants, and it means developing a sensitivity, an openness, an awareness of all kinds of weather patterns and patterns in nature.

GARY SNYDER, "THE INCREDIBLE SURVIVAL OF COYOTE"

Enlightenment science, following classical philosophy, organized life into three categories, in descending order of value: the rational, which was restricted to human beings; the animal, which was not rational but had sensory organs and mobility; and the vegetal. To describe a life as vegetal was to declare it devoid of any capacity for intelligence, activity, or sociality. That is why "vegetating" even now refers to monotonous time spent without social interaction or intellectual or physical activity. Brain-damaged humans who to all appearances are incapable of thought, motion, or communication are also colloquially called "human vegetables."[1]

Ecological thinking, however, defined mind as a phenomenon of complex feedback loops in which intelligence was manifested in physical action rather than limited to conscious thought. This redefinition of intelligence made it possible to see the responses of plants to their environments as a form of intelligence that functioned via electrical or chemical signals akin to the neural signals and bodily affects of humans. It suggested that plants were both intelligent and communicative.

This idea fostered various efforts to establish effective communications with plants. Cleve Backster and his followers used galvanometers to tap in to vegetal information networks by intercepting their electrical signals. The related practice of biofeedback therapy used a similar apparatus to tap into the autonomic nervous system, which was understood to be the seat of vegetal intelligence within the human organism. Leaders of the psychedelic counterculture and the holistic health movement advocated the ingestion of psychoactive plants in order to restore the natural balance between the conscious and unconscious subsystems of the human mind and enable telepathic communion with the environing "biofield."

The excitement generated by the possibility of talking to plants was matched, however, by an uneasiness about the effects of encountering their very alien subjectivity. In that respect, plants taken as intelligent beings were uncanny: their imperfect mirroring of human subjectivity was potentially threatening to it. The film *The Secret Life of Plants* drew its viewers in to dwell on this uncanniness, and to imagine what the modern world might seem like from a vegetal point of view. Additionally, certain styles of music—including sustained drones and Indian ragas—seemed to offer a middle ground where two such disparate intelligences as the vegetal and the human might meet.

Ecology and Plants' Rights

Disseminated through the environmentalist, ecofeminist, and American Indian movements, the ecological critique of Enlightenment science decried the dominant culture's denial of full personhood to all but the idealized white male subject, who epitomized Cartesian consciousness. But if ecological thinking advocated the extension of personhood to include all forms of intelligence, did this extend also to plants? It was not unthinkable. Rachel Carson in *Silent Spring* accused humans of a manipulative selfishness in our relationships with plants, using language that implied that such solipsism was tantamount to an ethical transgression:

> Our attitude toward plants is a singularly narrow one. If we see any immediate utility in a plant we foster it. If for any reason we find its presence undesirable or merely a matter of indifference, we may condemn it to destruction forthwith. . . . Many others are destroyed merely because they happen to be associates of the unwanted plants.

When Gary Snyder rallied the protestors at People's Park in Berkeley in 1969, he declared, "The most revolutionary consciousness is to be found among the oppressed classes—animals, trees, grass, air, water, earth." Theologian Kenneth Cauthen, who in the seventies formulated an ethics based on Alfred North Whitehead's process philosophy that he called "process ethics," wrote that any open system with a drive for self-preservation must be accorded some level of ethical consideration and some autonomy to pursue its own path and destiny—even a plant. He explained:

> It does seem clear that it matters to a flourishing, growing tree when it is cut down in a way that it does not matter to the stone when it is crushed. The tree as a whole dies. The organic system ceases to function, and it becomes a thing. . . . In this sense, at least, the tree is an experiencing subject.[2]

The possibility that plants might be persons entered American legal discourse in 1972, when Christopher Stone, a professor of law at the University of Southern California, published "Should Trees Have Standing? Toward Legal Rights for Natural Objects" in the *Southern California Law Review.* "Should Trees Have Standing" was written to weigh in on a case then before the Supreme Court, *Sierra Club v. Morton*. The Sierra Club had sought to prevent the Walt Disney Company from building a ski resort on Forest Service land in the Sierra Nevada Mountains. The federal district court had issued a preliminary injunction, but the court of appeals remanded it, on the basis that the Sierra Club had not demonstrated "standing"—that is, had not proven that as a legal entity it personally suffered injury from the proposed resort. The Sierra Club then appealed to the Supreme Court, which heard arguments in the case in November 1971. In his essay, Stone asserted that the issue of the Sierra Club's standing would be moot if, as he proposed, "we give legal rights to forests, oceans, rivers and other so-called 'natural objects' in the environment—indeed, to the natural environment as a whole."[3]

Awarding legal personhood to natural systems, Stone wrote, was necessary now that an ecological understanding of nature had allowed us to see them as self-organizing entities. "A few years ago the pollution of streams was thought of only as a problem of smelly, unsightly, unpotable water *i.e.*, to us," he gave as an example. "Now we are beginning to discover that pollution is a process that destroys wondrously subtle balances of life within the water, and as between the water and its banks."[4] The only way to protect such subtly organized systems from exploitation and destruction was by granting them legal standing.

Why shouldn't trees have standing, Stone asked, when a host of intangible and inanimate entities already held that legal right, including financial corporations and "the United States of America"—"creatures whose wants are far less verifiable, and even far more metaphysical in conception, than the wants of rivers, trees, and land." Legal personhood, he argued, has always been awarded based on a purely cultural distinction between persons and objects, which has changed over time. American history has witnessed the extension of legal standing to several classes of beings formerly treated as objects, including children, married women, and slaves. Now the concept of personhood should again be expanded, "to regard the Earth, as some have suggested, as one organism, of which Mankind is a functional part." To do this would change our conception of land use: instead of imagining the title holder as the "owner" of the land, who was free to do anything to it, he or she would be the land's legal "guardian," with rights and responsibilities toward it, as is the case for the legal guardians of children.[5]

Legal standing derives most of its practical meaning from the ability of the courts to award "damages" to a person whose rights have been infringed upon. To give trees standing would therefore imply seeing plants as capable of suffering, since the law reckons damages by assessing the pain and suffering caused by an injury. Stone did not balk at leading his readers across that threshold, however. "It is not easy to dismiss the idea of 'lower' life having consciousness and feeling pain, especially since it is so difficult to know what these terms mean even as applied to humans," he insisted. We legally recognize the existence of human pain and suffering "not because we can ascertain them as objective 'facts' about the universe, but because . . . [w]e come up with a better society by making rude estimates of them than by ignoring them." The same could be said of the suffering of forests.[6]

Striking a theme common to environmentalism of the 1970s, Stone hailed the idea of validating the suffering of plants as an expansion of our capacity for empathy. "We are cultivating the personal capacities *within us* to recognize more and more the ways in which nature—like the woman, the Black, the Indian and the Alien—is like us," he wrote. The cultural change that would be demanded by giving legal standing to trees would make us better human beings. It would help us overcome our psychological alienation from nature and from one another. He mused:

> If we only stop for a moment and look at the underlying human qualities that our present attitudes toward property and nature draw upon and reinforce, we have to be struck by how stultifying of our own personal growth and satisfaction they can become when they take rein of us. . . . What is it within us that gives us this need

> not just to satisfy basic biological wants, but to extend our wills over things, to object-ify them, to make them ours, to manipulate them, to keep them at a psychic distance? . . . Should we not be suspect of such needs within us, cautious as to why we wish to gratify them?

It would be better for us, he concluded, "to get away from the view that Nature is a collection of useful senseless objects," and to be able to see ourselves as part of a larger natural system governed by give-and-take. Granting trees standing would give human beings an opportunity to develop "our abilities to love . . . to reach a heightened awareness of our own, and others' capacity in their mutual interplay."[7]

Stone's impassioned plea did not go unheeded. Later that year, the Supreme Court narrowly (4–3) upheld the ruling of the appeals court dismissing the Sierra Club's suit. But in a dissenting opinion, Justice William O. Douglas cited Stone's essay and asserted that "concern for protecting nature's ecological equilibrium should lead to the conferral of standing upon environmental objects to sue for their own preservation."[8]

In his opinion, Justice Douglas assigned an imaginary voice to a hypothetical river ecosystem, writing:

> The river as plaintiff speaks for the ecological unit of life that is part of it. . . . The voice of the inanimate object, therefore, should not be stilled. . . . Then there will be assurances that all of the forms of life which it represents will stand before the court—the pileated woodpecker as well as the coyote and bear, the lemmings as well as the trout in the streams.

For Douglas, such language remained a rhetorical conceit; "those people who have so frequented the place as to know its values and wonders will be able to speak for the entire ecological community," he wrote. In his essay, however, Stone suggested the more radical possibility that a human might speak for the trees not as their representative, but as their interpreter. In this he was closer in sensibility to Snyder, who imagined the shaman-poet as nature's spokesperson, delivering "a speech on the floor of Congress from a whale." "Natural objects *can* communicate their wants (needs) to us, and in ways that are not terribly ambiguous," Stone insisted. Plants communicated with humans through our sense of touch: "The lawn tells me that it wants water by a certain dryness of the blades and soil—immediately obvious to the touch— . . . and a lack of springiness when walked on."[9]

The belief that plants could "tell" people things if they were properly attuned to them so that plant and human formed a single intelligent system was shared by many ecological thinkers. With this understanding, a variety of methods were developed for tuning in to the voices of plants. While Stone relied on an unaided sense of touch, L. Ron Hubbard and Cleve Backster used galvanometers to "listen in" on plants' electrical impulses. Backster was surprised to find, when he did so, that the plants' communications were mainly concerned with listening in on him! Marcel Vogel and Randall Fontes accepted this idea and adapted Backster's technology to bring plant intelligence to bear as a means for gauging and healing the human psyche; this possibility underlay the invention of the biofield sensor as well as of biofeedback therapy. Taking a different approach, other thinkers—from Gary Snyder and Andrew Weil to Stephen Gaskin and the McKenna brothers, Terence and Dennis—preferred to rely on intuition rather than on scientific gadgetry. They believed that plants came equipped with their own technology for making their voices heard: in the molecular structure of chemicals found in certain psychoactive plants, which, when linked to the neurons of the central nervous system, enabled humans to tune in telepathically to the continual conversation among plants and animals that was going on all around them.

Vegetal Signings

The galvanometer is an instrument invented near the beginning of the nineteenth century, which uses a magnetic needle to detect electrical currents. It was named after Luigi Galvani, who in 1791 discovered (using a dead frog) that electrical impulses are the means by which the nervous system communicates with the rest of the body. This gave rise, in the early twentieth century, to the use of the galvanometer in the quasi-scientific field of psychophysiology, which explored the possibility of interpreting physiological measurements as indicators of inner emotional states.

The most widespread example of psychophysiology in the mid-twentieth century was the lie detector machine. The polygraph had been created in the late nineteenth century as a medical apparatus that could conveniently monitor a variety of patient symptoms simultaneously, including skin conductivity (using a galvanometer), depth and rapidity of breathing, pulse rate, and blood pressure. In the early 1930s, psychophysiologist William Moulton Marston repurposed the polygraph as a "lie detector machine" and offered his services to de-

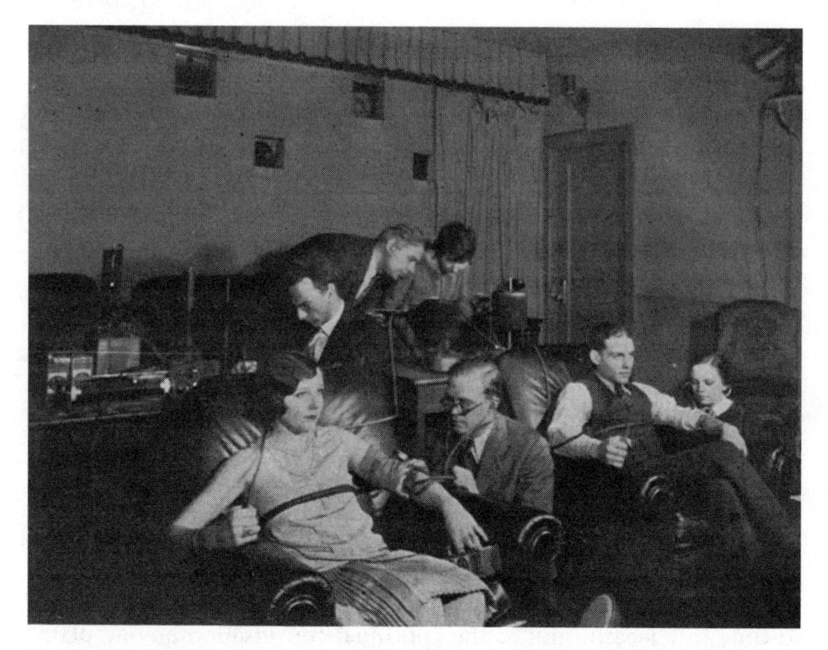

Figure 5. William Moulton Marston (*back left*) uses the polygraph in pre-screenings for Universal Studios. (The galvanometer is held in the viewer's right hand.)

partments of law enforcement.[10] To accomplish this repurposing, he interpreted the changes that the machine detected in the various bodily processes it measured, attributing them psychophysiologically to an inner emotional state that was hypothetically produced by lying. Over subsequent decades, different investigators identified this emotion variously as guilt, fear, and even gloating. Before branding his apparatus as a lie detector, however, Marston had marketed it to Hollywood executives as a tool for gauging people's emotional responses to movies (fig. 5). In that context, it purported to measure excitement. In the final analysis, then, psychophysiologists could not say exactly what emotional state a change in skin conductivity might indicate; but they were sure that it meant something.

Another mid-century psychophysiologist who used the galvanic skin response (GSR) as an index of emotional excitement was L. Ron Hubbard, founder of the religion of Scientology. Hubbard originally developed his religion as a branch of psychology that he called Dianetics. Beginning in 1951, he employed the galvanometer—which he called the "electro-psychometer," or "e-meter" for short—in a process known as "auditing." The purpose of auditing was to "dig out" buried memories that were blocking the subject's psychological or spiritual progress, in order to "clear" them of emotional energy. If the e-meter's needle swung to the right during questioning, it indicated a "charged" emotional state that was connected to a memory that had not yet been "cleared."[11]

Although it had been known since 1901 that plants responded electrically to stimuli, it was not until psychophysiologists had conceptualized a link between electrical conductivity and emotions that this phenomenon was interpreted to indicate the presence of emotions in plants. Hubbard applied his e-meter to tomato plants and geraniums while living in England in 1959 (fig. 6). For the benefit of a reporter from the London-based *Garden News*, he attached a geranium to an e-meter and tore off a leaf; the needle jumped. To Hubbard this

Figure 6. L. Ron Hubbard applies an electro-psychometer to a tomato plant. Photo by Scott Lauder.

demonstrated that plants had at least rudimentary emotions, among them a "fear of death." The journalist from *Garden News* agreed and published a story titled, "Plants Do Worry and Feel Pain," which attracted the attention of Alan Whicker of BBC television's *Tonight* program. *Tonight,* which had approximately seven million viewers, aired Whicker's interview with Hubbard on the topic in December 1959.[12]

Beginning in February 1966, lie-detector expert Cleve Backster also used a galvanometer to measure emotional responses in plants. Backster's work became widely known after it inspired Peter Tompkins and Christopher Bird to write their 1973 national best seller, *The Secret Life of Plants.* Backster's findings were first published in late 1968, in the obscure *International Journal of Parapsychology,* but he gained mainstream media attention when the environmentalist magazine *National Wildlife* published a feature article on his work in February 1969 under the title "The Man Who Reads Nature's Secret Signals." *National Wildlife*'s editor, Dick Kirkpatrick, appended a note to the article in which he vouched for Backster's credibility and attested to witnessing personally several instances of plants' responding to a variety of psychic and physiological stimuli. The magazine also published, as a sidebar, a photo of the strip of polygraph tape on which was recorded "Kirkpatrick's interview with Backster and a philodendron"—with annotations to indicate what the philodendron was responding to each time the galvanometer registered a reading.[13]

The *Wall Street Journal* picked up on the *National Wildlife* story, and in February 1972 reported on Backster's work with an edge of bemusement but an overall favorable tone. This exposure led to a round of articles in the daily press. The Santa Rosa (CA) *Press Democrat,* for instance, printed a feature article in September 1972 titled "Talk Nice to Plants . . . They May Be Listening!" The writer explained that the secret to a green thumb was treating plants with "tender loving care" and warned, "If a person thinks derogatory thoughts about one of his plants or talks to a plant in a derogatory way, while praising some other plant, the first plant can be made to wither to the point of death." Tompkins and Bird (who were parapsychologists, not botanists—Tompkins had previously written a book titled *Secrets of the Great Pyramid,* and Bird would go on to write about dowsing) coauthored an enthusiastic article that appeared in the November 1972 issue of *Harper's* magazine, in which they claimed exaggeratedly that "scientists everywhere in the world, amazed by the results of careful laboratory experiments, find themselves coming to the conclusion that plants have emotions similar to those of human beings, [and] that they respond to affection."

In their best-selling book, Tompkins and Bird went into greater detail, cataloging Backster's repeated successes in communicating with plants via the galvanometer.[14]

Even those who did not buy the book absorbed its message through the popular press, television, or the movies. In 1973 Tompkins and Bird were interviewed by Barbara Walters on NBC's *Today* show and by the *CBS Morning News* on the day after Christmas.[15] It was not long before thousands of people began talking to their plants, and plant nurseries saw an upsurge in popularity for potted philodendra, which were the plants favored by Backster in his experiments. In 1979 a film adaptation of *The Secret Life of Plants* was released, featuring footage of Backster in his laboratory, with an original soundtrack by Motown legend Stevie Wonder.

Before becoming famous in the seventies as the man who tuned in to plants, Backster had logged considerable experience interpreting galvanometric responses as a form of unconscious communication. In 1948 he was trained by the CIA in the use of the polygraph for interrogation in conjunction with hypnosis and sodium pentothal. He then took over as head of the Keeler Polygraph Institute in Chicago, where late one night in the winter of 1966 he first connected the potted plant in his office to a galvanometer by one of its leaves and tried to devise ways of getting a reaction. He thought of burning one of its leaves, and as he had the idea, the galvanometer registered a sudden change in the plant's surface conductivity—akin, Backster later wrote, to that which lie-detection experts consider "typical of a human subject who might have been briefly experiencing a fear."[16]

Backster believed that this experiment and the many others that he subsequently conducted demonstrated that plants were sensitive to the energy—even to the thought waves—of the life-forms around them. Since they possessed no identifiable sense organs to gather and process this information, he attributed their knowledge to a mysterious form of receptivity and communication that he called "primary perception." As Tompkins and Bird explained in *The Secret Life of Plants*, Backster "hypothesize[d] that the five senses in humans might be limiting factors overlying a more 'primary perception,' possibly common to all nature. 'Maybe plants see better *without* eyes,' Backster surmised: 'better than humans do with them.'" Whereas the nervous system processed and packaged humans' bodily perceptions, even to the extent of filtering out the bulk of unconscious awareness, plants absorbed and responded to all ambient energy. As a result, *The Secret Life of Plants* reported, they were "able to perceive and to react to what is happening in their environment at a level of sophistication far surpassing that of humans."[17]

According to Backster, plants were aware of the well-being of other plants in their vicinity, but they were particularly "attuned" to the presence and disposition of nearby animals. He hypothesized that this was because animals, being mobile, presented the greater threat. Plants could become sensitized and "attuned" to animals that frequented their local territory. Then their awareness of that animal's feelings persisted even when the animal was miles distant.[18]

Recognizing that plants' relations with animals were not always antagonistic but often symbiotic, Backster further hypothesized that plants read animal emotions also as a way of extending the scope of their sensory perception. "When other life forms were threatened, the plant apparently perceived its wellbeing as threatened" as well, he wrote. The film version of *The Secret Life of Plants* chronicled one of Backster's experiments in this vein. In the film, as Backster wanders distant streets so as not to affect the experiment's outcome with his own emotions, a mechanical apparatus at a random instant dumps living brine shrimp into boiling water so they die; the plant's "distress" response is shown to register on the galvanometer. Backster comments, "I think it's the smallness of the event that makes it so significant. It means that even on the lower levels of life, there is a profound consciousness or an awareness that binds all things together."[19]

Backster also discovered that under acute emotional stress, a plant would exhibit a "fainting" response and cease reacting to all subsequent stimuli. He hypothesized that this was caused by an anticipation of imminent destruction and was a response that was triggered in the field in the final moments before a plant was eaten. Tompkins and Bird suggested that this was another form of symbiosis that had developed through coevolution. "Plants and succulent fruits might *wish* to be eaten, but only in a sort of loving ritual, with a real communication between the eater and the eaten," they wrote. Plant subjectivity seemed to include a strategy of cooperation with other organic subsystems that placed scant emphasis on the survival of any individual plant. Fainting was the apt vegetal alternative to the animal strategies of fight or flight.[20]

Backster was eager to have his ideas validated by the scientific community, and he tried to conform to the requirements of the scientific method by introducing an increasing number of controls into his experiments. He placed the plant and the investigator in separate rooms; he put the plant behind a sealed partition and in a copper cage; he devised a remote timer so that he could absent himself entirely at the moment of stimulus, lest his expectations affect the outcome of the experiment. The film *The Secret Life of Plants* took Backster's part in this struggle for scientific legitimacy. From its depictions of his automated laboratory—the silent philodendron attached to a galvanometer; the brine shrimp mechanically suspended over boiling water; the pen of the polygraph scrolling amidst a bank of sophisticated electronic equipment—the viewer is meant to infer an atmosphere of bona fide science. But scientists publishing in reputable journals announced that they could not replicate Backster's results.[21]

Backster himself acknowledged that reproducibility was a problem; but he did not see this as invalidating his conclusions. There was, first of all, the fainting response: it was impossible to predict when a vegetal subject would simply stop reacting to provocations. Even more problematically, the intelligence of plants was dismissive of the laboratory setup. The artificiality of its stimuli diminished their responsiveness. "The plants seemed to know when you didn't mean it," he wrote; after three or four repetitions under the same conditions, they typically ceased reacting. To Backster, this demonstrated not that they were unintelligent but, conversely, that they were learning. Finally, the emotions of an investigator intent on demonstrating that plants were insentient was sure to dampen their responsiveness. This was an epistemological problem that plagued the scientific method in studies of involuntary intelligent subjects generally: what if they chose not to do something, even if they could?[22]

One person who claimed that he had been able to reproduce Backster's results was Marcel Vogel, a researcher on magnetics and phosphor luminescence at the IBM laboratories in California. Interviewed by Tompkins and Bird, Vogel came to Backster's defense, insisting that scientists could not replicate Backster's results "if they approach the experimentation in a mechanistic way . . . and don't enter into mutual communication with their plants and treat them as friends." The problem lay not with Backster's conclusions but with the scientific method and its insistence on the myth of the objective observer. Studying this kind of phenomena, Vogel maintained, required the exactly opposite approach: "The empathy between plant and human is the *key*. . . . But this runs counter to the philosophy of many scientists, who do not realize that creative experimentation means that the *experimenters must become part of their experiments*."[23]

Like Backster, Vogel had long been interested in hypnosis—a background to which he attributed his ability to reproduce Backster's experiments while others couldn't, since establishing empathic rapport with a plant demanded a quality of attention akin to that of mesmerism. Tompkins and Bird described Vogel's attuning himself to a philodendron attached to a galvanometer in the spring of 1971:

> Vogel stood before the plant, completely relaxed, breathing deeply and almost touching it with outspread fingers. At the same time, he began to shower the plant with the same kind of affectionate emotion he would flow to a friend. Each time he did this, a series of ascending oscillations was described on the chart by the pen holder. At the same time Vogel could tangibly feel, on the palms of his hands, an outpouring from the plant of some sort of energy.

Plant intelligence, Vogel concluded, had evolved to function as part of a "circuit" including other plants and animals. Together they formed a communications network. Connected empathically, "man and plant seemed to interact, and, as a unit, pick up sensations from events, or third parties."[24]

Randall Fontes took over Vogel's experiments in 1972. Previously, he had been a hippie and a follower of Hindu Swami Muktananda. Fontes referred to Vogel's networked intelligence of plants and animals as the "biofield" and associated it with the Hindu concept of *shakti*. He hoped to develop some practical applications for the galvanometric approach, as he explained to a local journalist: "People have various levels of consciousness, many of them hidden, and through the use of highly sensitive plants . . . these levels could be measured." Plants' responses to people, specifically because they were *not* objective, could provide a sophisticated form of aptitude testing.[25]

The idea that plants formed part of a networked, systemic intelligence sounded promising enough that both the CIA and U.S. Army Intelligence explored the possibility of leveraging "biofield" communications as a Cold War weapon. Through the Stanford Research Institute, Fontes received funding in 1975 from the CIA's "Stargate" program to test whether a mimosa plant could be mobilized as an "organic biofield sensor." The objective of this research was to determine whether plants could be used as a tool of remote covert surveillance of a person's emotional state.[26]

As reproduced for the film *The Secret Life of Plants*, Fontes's biofield sensor consisted of a mimosa plant sequestered behind glass and wire cages (to prevent electronic, mechanical, or chemical interference) and attached to a galvanometer. A human subject approximately five feet away was hooked up to a separate galvanometer, in order to ascer-

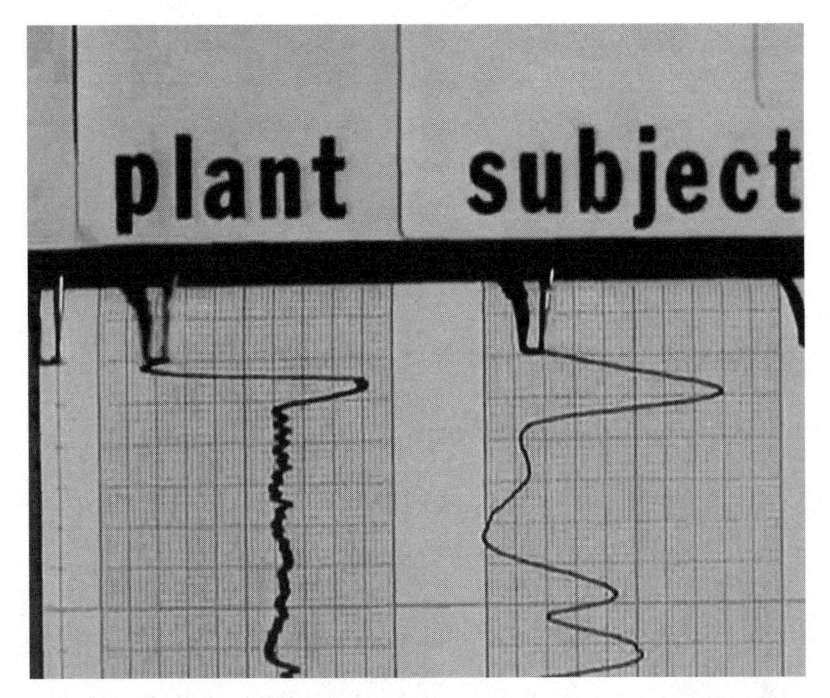

Figure 7. A mimosa plant responds to human sexual excitement in *The Secret Life of Plants*.

tain whether his emotional responses would be echoed by the plant. The human subject watched a filmstrip for emotional stimulus, and the viewers of *The Secret Life of Plants* watch it along with him. Its images of trees and of children on bicycles abruptly segue to an extreme close-up of a woman splashing water on her breasts in the shower, and the recording pens of both galvanometers are shown to move in concert (fig. 7).

Fontes's final report for the CIA-funded project pronounced the biofield sensor a failure. He could achieve only a 20 percent correlation between plant and human responses. Nevertheless, he insisted, this was still a much higher correlation than could be accounted for by mere chance. He later set up a biofield sensor at Ripley's Museum on Fisherman's Wharf in San Francisco; there, the *San Francisco Examiner* reported, "Visitors can try directing nasty thoughts at the plant and observe the results, if any."[27]

By 1977 Fontes and his partner Bob Swanson had achieved the status of minor celebrities by appearing on television specials and talk shows (including *The Tonight Show Starring Johnny Carson*) as evangelists of the word that "plants act like an antenna for raw psychic energy." Perhaps as a result of the influence of Reichian psychology in sixties countercultural circles, the biofield was thought to be particularly attuned to sexual energy. *The Secret Life of Plants* showed Fontes's mimosa picking up on lust in particular. Tompkins and Bird reported similar incidents occurring in Vogel's laboratory. Vogel explicitly advocated affective interactions with plants as a therapy for overcoming "armoring"—a socially pathological condition, according to Reichians, in which the individual was psychologically, sensorily, and sexually walled off from others.[28]

Listening to Our Vegetal Selves

Consistent with its discursive privileging of sensuality as a means to knowledge, ecological thinking advocated a revision of the power dynamics structuring relations between the conscious and the unconscious mind. This was in keeping with its critique of centralization and centralized authority structures, since the conscious mind was understood to be the locus of a centralized authority in the self. One means of fostering these revised dynamics entailed using a galvanometer, in a variation on Backster's techniques, to improve communications to consciousness from the autonomic nervous system, which was equated with the "vegetal mind" within each of us.[29]

Enlightenment science divided the human nervous system into two parts. The central nervous system, located in the brain and spinal cord, controls willed, voluntary movements. The autonomic nervous system, organized by a chain of ganglia running down either side of the spinal cord, produces involuntary affective and visceral responses. This ganglionic nervous system, discovered by French neuroanatomist François Xavier Bichat at the end of the eighteenth century, was named the "vegetative nervous system" by German physician Johann Christian Reil. It was later (circa 1900) renamed the autonomic nervous system, in order to emphasize that it was not controlled by the will. Enlightenment science considered its responses inferior to the actions of the central nervous system under rational control.[30]

General systems theory broke down this dichotomous view of the "two" nervous systems. At the 1968 Alpbach Symposium organized by Arthur Koestler, psychologist Jerome Bruner asserted that the central nervous system and autonomic nervous system were two subsystems connected by complex feedback loops. With this understanding, the instruments of psychophysiology were repurposed as the tools of biofeedback.

The term "biofeedback" was coined in 1969 at a conference in Santa Monica, California, which saw the founding of the Biofeedback Research Society. The aim of biofeedback therapy was to improve communications between the central and the autonomic nervous systems by using the galvanometer to amplify the signals of unconscious bodily processes, so that they could be used to modify conscious thought patterns. Although the practitioners of biofeedback sometimes resorted to the use of the polygraph, they most often made use of the newer technology of the electroencephalogram, or EEG. (As his fame grew, Cleve Backster was also able to supplement his studies of "primary perception" using the more sophisticated EEG technology.) The EEG has multiple electrodes connected to a single galvanometer. These electrodes are attached to the outside of the scalp to detect fluctuations in the voltage of the local electrical field caused by synaptic activity in the brain. Those periodic fluctuations create what are known as "brain waves."[31] Practitioners of biofeedback therapy attempted to alter their mind-set, guided by such indices as their success in slowing their heart rate, lowering their blood pressure, and producing particular brain wave frequencies: the alpha wave, associated with a state of restful relaxation and openness; and the theta wave, associated with a meditative or mesmeric state.[32]

The attendees at the Santa Monica conference of the Biofeedback Research Society retroactively recognized the e-meter of Dianetics as a primitive form of biofeedback therapy. But in her 1974 book *New Mind, New Body: Bio-Feedback*, UCLA-based psychopharmacologist Barbara Brown, who was the conference organizer, contrasted biofeedback's use of the galvanometric skin response to that of the lie detector. While the lie detector used the galvanometer to create a "peep show" that aimed to make our own bodies "give us away," biofeedback concentrated on "the great potential benefits that can derive from intelligent use of the skin's emotional speaker . . . a more direct route to the subconscious than conventional psychotherapy."[33]

According to Brown, biofeedback therapy inverted older psychological models that attempted to address neuroses by bringing the consciousness to bear on their unconscious sources. Like Gregory Bateson, she saw the unconscious as the source of a more complete, holistic wisdom than consciousness:

Popular notions of psychiatry have given us to understand that emotional problems arise from conflicts between conscious and subconscious feelings, and in general it is the subconscious which is held to be the offender. . . . Yet if we examine carefully the thousand bits of psychophysiologic evidence that reveal subconscious judgments to be sound and reasonable, it makes one wonder whether or not we have been fighting the emotional battle without knowing which side we're on.

She asked her readers, "Are we now ready to be Alice in Wonderland and look at the world of conscious constraints through the eyes and ears and skin of a wise and prudent subconscious?" From that perspective, neurosis was reimagined as a message from the individual's whole psychophysiological network, demanding a change in lifestyle by somatically "crying out warning signals to a consciousness" that was not listening.[34]

Biofeedback therapy enlisted the autonomic nervous system as a force for holistic healing. In that respect, it was part of a larger cultural movement. Holistic physician Andrew Weil wrote in his book *The Natural Mind*, published in 1972, that better communications between the autonomic nervous system and the central nervous system were necessary to both psychological and physical health. Improved communications could be fostered through biofeedback, he acknowledged, but he preferred to cultivate them using a variety of more natural practices that enabled the autonomic nervous system to assert itself while the "observing ego gets out of the way." These included yoga, meditation, laughing, crying (Weil described "psychogenic tearing" as a literal "spillover of autonomic activity"), and even vomiting ("You can learn to do it so it's not painful, not forceful"). "The autonomic nervous system seems to have very direct connections to every kind of consciousness but the ordinary one of the ego," he summarized. "This channel between the mind and body is wide open whenever we are in an altered state of consciousness that focuses our awareness on something other than our ego and intellect."[35]

In keeping with this logic, Weil also advocated the use of psychoactive plants like marijuana to cultivate a more balanced psychological orientation. He believed that the altered state of consciousness they fostered would be conducive to a more beneficial relationship of coexistence with nature. It was, in his eyes, conventional medicine's dependency on antibiotics to fight an inevitably losing battle against germs that constituted American society's real "drug problem."[36]

Holistic medicine pursued the ecological critique of Enlightenment science into the medical realm. Weil criticized allopathic medicine (the approach certified by the American Medical Association) and its germ theory of disease as myopically mechanistic. These focused on specific pathogens, rather than looking at all the relations constituting the system comprised of the ill person in his or her environmental context. Weil wrote that instead of demonizing and attempting to eradicate other organisms that were in any case constantly coevolving with us, "it's important to see germs as agents rather than causes of disease." He argued:

> We live in a world full of germs, some of which are correlated with physical symptoms of infectious disease. But only some of us get infectious diseases some of the time. Why? Because there are factors *in us* that determine what kind of a relationship we will have with those germs that are always out there—a relationship of balanced coexistence or one of unbalanced antagonism.

Therefore he considered the use of psychedelic drugs as a means to holistic health. "I think that getting high in a good way is an essential part of learning to control disease," he told an interviewer. "When you're high you are using your mind in an expanded way, and it's through such expanded consciousness that you learn to run your body smoothly."[37]

Psychedelics: Talking Plant-to-Plant

For many ecological thinkers of the seventies, psychedelic drugs provided a kind of interface between the human mind and plant intelligence. Andrew Weil emphasized the coevolutionary development of intelligence in humans and some plant species. "It appears that some plants make some kind of evolutionary decision to become involved with our trip. Marijuana is very involved," he told an interviewer in 1975.[38] While biofeedback was analogous to talking to the plant within (the autonomic nervous system) by means of a galvanometer, taking psychedelics was talking to the plant within, by means of ingesting plants.

Psychedelics seemed to indicate that the intelligence of the autonomic nervous system was plant-like in its gravitation toward a collective self, while the conscious mind was individuated and competitive. As an example of "the tendency of the ego to focus on the differences rather than the similarities between itself and things out there," Weil described how psychedelics had changed his attitude toward bees:

> Like many of my friends, I projected my sense of the hostility of nature onto certain insects, and while my fear of them did not approach phobic proportions, it was sufficient to keep me from relaxing completely in a wild setting. Although I did not understand it at the time, these feelings arose entirely from my conceiving of these insects (particularly bees and wasps) as fundamentally different from myself and, "therefore," able to harm me. Two years ago, during an LSD trip, I found myself extremely high and unattached to my ego in a field with many bees. For the first time in my life I experienced these creatures as essentially similar to myself and was able to see in them extraordinary beauty I had never before noticed. Since that time, I have learned to extend that feeling to most other insects, many of which I now regard as friends and sources of pleasure. Especially interesting is my finding that the insects themselves appear to behave differently toward me. . . . I have never been stung and appear to cause them no discomfort or alarm. Needless to say, this change (which had its origin in an altered state of consciousness triggered by a drug) has been a source of great joy.[39]

Weil condemned the illegalization of drugs along the same lines that he criticized efforts to eradicate germs. For him, cultural attitudes toward psychedelic drugs were a subset of the broader issue of plant-human relations. Psychoactive plants were not the problem; our relationship to them was. "You can try to wipe drugs out, or you can work to improve relations with them. It's the same [as] with germs," he insisted. Some traditional Native American cultures, Weil noted, had institutionalized symbiotic relations with psychedelic plants by ritualizing their use. His book *The Marriage of the Sun and Moon* consists of essays written between 1971 and 1977 on Native American relations with a wide range of psychoactive plants—including marijuana, coffee, mangos, chilies, mushrooms, ayahuasca (yagé), and coca. In it, he also promoted eating only unrefined sugar and whole grains. In this respect, Weil's thinking illuminates the historical connections between the psychedelic counterculture, holistic healing, and the origins of the health food movement. As he summarized, "More and more people today are exploring their environment and learning to use what nature provides" us through plants.[40]

Fungi, although technically not plants, were included in the counterculture's project of restructuring plant-human relations. Weil included chapters on mushrooms in *The Marriage of the Sun and Moon*. The film version of *The Secret Life of Plants* incorporates numerous sequences of time-lapse photography in which mushrooms and other fungi seem to grow in fast-forward. The fungi of greatest interest to Weil were hallucinogenic mushrooms of the genus *Psilocybe*.[41] But it was also often mentioned that LSD was a derivative of ergot.

Many of the people at the forefront of ecological thinking in the seventies had passed through the fungal doors of perception in the sixties. Biofeedback expert Barbara Brown had previously been involved in LSD research and on expeditions to Mexico in search of "magic mushrooms." Lynn Margulis ingested acid in 1963. Randall Fontes attributed his intellectual breakthrough to LSD, as did Fritjof Capra and the dolphin expert John Lilly. Gregory Bateson wrote that his experience with LSD helped him to imagine what it would mean to have a systems concept of self as an everyday habit of thought. Dancer and choreographer Simone Forti wrote in 1974 of the contributions that LSD had made to her perception of "flow," which she described in a vocabulary drawn from systems theory:

> All the acid I took seemed to break down the barriers to perception and communication between the myriad systems and processes which house the self and within which my own identity lived as one interpretation among many others with which I coexisted as in a fertile jungle of interpenetration of life.

Ethnobotanist Terence McKenna wrote in his journal from 1971 that "the mushroom, as is said of peyote, teaches the right way to live."[42]

The most complete theorization of how psychedelics enabled talking with plants came from the brothers Dennis and Terence McKenna, who in 1971 traveled to Colombia to investigate the shamanic use of psilocybin mushrooms and ayahuasca ingested in combination. "The mushroom must be heard," Dennis proclaimed. The brothers became convinced that Gaia was talking to them using the mushroom as a medium. Terence hypothesized that mushrooms were "a kind of intelligent entity—not of this earth," which interacted with earth animals' brain chemistry, catalyzing communications among the species. Dennis later recounted, "We did have a sense of being in the presence of an 'other,' an entity of some kind that was fully participating in the conversation, though in a nonverbal or perhaps metalinguistic way [by affecting their choice of words]. We came to think of this other as 'the Teacher.'"[43]

The McKennas made much of the fact that mushrooms are only the visible fruit of underground mycelial webs analogous to neural networks. Writing under the pseudonyms O. T. Oss and O. N. Oeric, they let "the mushroom speak" on that point in the introduction to their

1976 book, *Psilocybin: Magic Mushroom Grower's Guide*: "My true body is a fine network of fibers growing through the soil. These networks may cover acres and may have far more connections than the number in a human brain. My mycelial network is nearly immortal." Furthermore, they represented the mushroom as saying that mycelial networks were seeded across the galaxy and in constant communication with one another. Spores spread this network from planet to planet, so that it could make contact with local species in search of other intelligences with which it could become linked. "The mycelial body is as fragile as a spider's web but the collective hypermind and memory is a vast historical archive of the career of evolving intelligence on many worlds." Humanity was now, the McKennas' mushroom foretold, "on the brink of the formation of a symbiotic relationship with my genetic material that will eventually carry humanity and earth into the galactic mainstream of the higher civilizations."[44]

If mushrooms, according to the McKennas, were "loquacious" appendages of an alien intelligence, on Earth to facilitate "a planet-saving shift of consciousness," it was the ayahuasca plant that hummed with the indigenous voice of the rain forest. Ingesting yagé in conjunction with the mushrooms put the McKennas, Terence wrote, "in touch with the living mind of the tropical forest." He had a sensation of "the suspension of time, of turning and turning in a widening green world that was strangely and almost erotically alive . . . the jungle as mind, the world hanging in space as mind—images of order and sentient organization came crowding in on all sides."[45]

Searching afterward for an intellectual grasp of their psychedelic experiences, the McKenna brothers turned to Alfred North Whitehead's metaphysics and Ludwig von Bertalanffy's general systems theory. When they came to write about how yagé could transmit the voice of the rain forest, they explained it using the language of cybernetics. "Perhaps tryptamine compounds [the psychoactive ingredient in yagé] are the mediators of the signaling mechanisms of the command-and-control structure that regulates and integrates whole ecosystems," Terence suggested. The result of ingesting tryptamine, therefore, Dennis wrote, was to gain access to the telepathic communications constantly passing through the ecosystem. It gave the user "access to an enormous, cybernetically stored fund of information" that existed in nature as a "planet-wide phenomenon."[46]

Listening and speaking to plants through the use of psychedelics could only be accomplished via telepathy. Telepathy, as it was understood by both Andrew Weil and the McKenna brothers, was an effect of networked intelligences in an environmental context, like Marcel Vogel's empathic attunement. As such, it was consistent with such phenomena as "animal magnetism," "group mind," and murmuration.[47] Scientific studies of telepathy as a paranormal phenomenon approached it wrongly by decontextualizing it. Its existence was not demonstrated by the percentage success rate of two subjects attempting to send and receive intentional, arbitrary messages like a shape or a word printed on a card, from one consciousness to another. Instead, telepathy involved tapping unconsciously into the environing biofield's ongoing collective awareness of its shared situation.

According to Weil and the McKennas, psilocybin, ayahuasca, and peyote all facilitated telepathy by toppling the conscious ego and catalyzing the networking of intelligence. Terence described his brother and himself as merging under the influence of psilocybin in May 1971 into a single "cybernetic unit." For his part, Weil wrote, "I have no doubt that telepathy exists. In fact, I think it is so common that we do it all the time without being aware of it or without attaching significance to it."

> That communication of this sort exists is really no more remarkable than the fact that intuition exists. . . . Telepathy is nothing other than thinking the same thoughts at the same time others are thinking them—something all of us are doing all the time at a level of our unconscious experience most of us are not aware of. Become aware of it and you become telepathic automatically. . . . And perhaps the people best qualified to teach us about this system are not neurophysiologists but Indians who regularly go off into the forests to see the same visions simultaneously.[48]

The Peruvian Indians, he noted, used ayahuasca ritually to facilitate their access to a collective unconscious; it was they and not scientific investigators who could tell us the truth about telepathy.

Stephen Gaskin was another figure of the seventies counterculture who claimed that psychedelic drugs could produce telepathic connections by increasing one's environmental awareness. A leading figure among the hippies of Haight-Ashbury who in 1970 led his followers cross-country by caravan to found a rural commune in Tennessee, Gaskin wrote in his memoir, *Haight Ashbury Flashbacks*, that "Dr. David Smith, of the Haight Street Free Clinic, said, 'Acid lowers your powers of discrimination until everything seems important.' When I heard that, I said, 'No. Acid raises your powers of integration until everything is important.'" For this reason, he continued, psychedelics enabled one to become more aware of natural omens and had helped Native

American tribes to make decisions by silent consensus. Psychedelics refocused one's attention away from the abstractions insisted on by verbal language, to the quick and "complicated energy exchanges" that constituted subtle nonverbal communications.[49]

Gaskin phrased many of his ideas in ecological terms. Telepathy exists, he wrote, because individuals are all parts of a single system: "Just because something is separated by a surface does not mean it is two things." Under the influence of psychedelics, he had ecological visions that made him progressively more environmentally conscious. He recounted for his readers one in which "it was like the world was a completely round ball, like an ocean with waves and ripples. And when you got close enough to see it, the waves and ripples were people, genes." This led him to a visceral comprehension of

> the oneness of life. Not just man, primates, slow loris, but the trees, and the grass, and the monkeys and the mastodons and everybody. We were really all one thing, and any violence done to any other was violence done to life; and as long as we considered the greatest good of mankind as the greatest good, we could destroy our planet.

On the basis of this insight, he formulated an environmental ethic that he communicated to his followers: "Continue to widen your circle of acquaintances until it includes all sentient beings."[50]

Freaks Like Us

According to these thinkers, plants possessed an ambient and yet alien intelligence that it was both necessary and difficult for humans to fathom. Another veteran of the sixties counterculture who like Gaskin made his way to a rural commune was Stephen Diamond, formerly of the Liberation News Service. Diamond's memoir of communal living, *What the Trees Said*, was published in 1971. In it, he recounted his experience of ingesting mescaline (the synthetic form of the psychoactive chemical in peyote) and walking down the driveway toward a line of maples. From a distance he felt confronted by their energy. On getting closer, he experienced "what seemed to be a bolt of vibrational energy coming from the trunk of the first tree . . . the tree and I were having a conversation, but no words or even picture images had appeared." He was both terrified and elated to acknowledge the tree as an "ancestral relative," and felt "a terrible joy and sorrow" at reconnecting with his

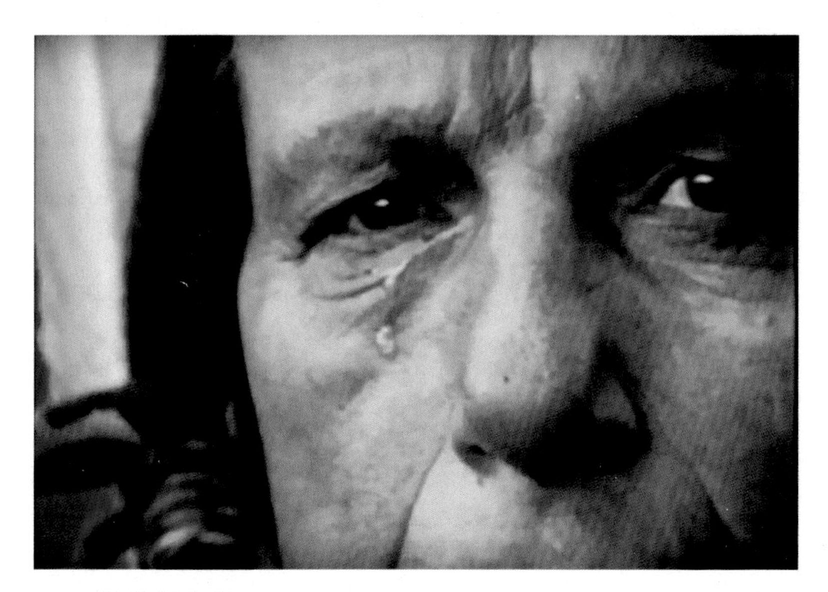

Plate 1. The Crying Indian.

Plate 2. Macro- and time-lapse photography show slime mold dancing uncannily in *The Secret Life of Plants*.

Plate 3. A seed floats like a planet in space in *The Secret Life of Plants*.

Plate 4. *Noise!* PSA, 1977.

Plate 5. "Quiet!" from *Noise!* PSA, 1977.

Plate 6. Robert Redford with Let's Merge in *The Electric Horseman*.

Plate 7. Three still shots from Carroll Ballard's *Rodeo*.

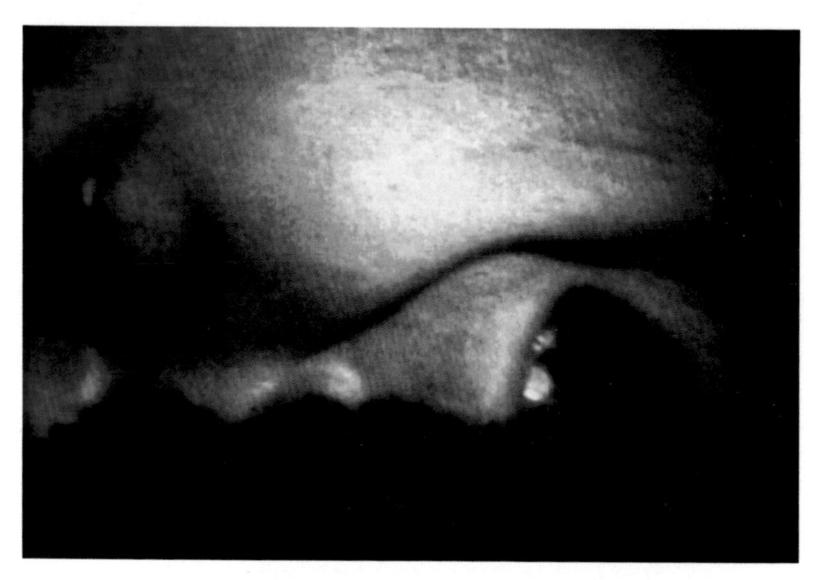

Plate 8. Close-up of the cowboy's open mouth after he is thrown by the bull in *Rodeo*.

long-sundered kin. But "what the trees said," though this was the title of his memoir, was ultimately untranslatable into words, and left him shaken by a sense of the uncanny.[51]

The uncanny (in German, *Unheimliche*) is an uncomfortable sensation that occurs when something is both familiar (*heimliche*, or homelike) and yet somehow at the same time alien. Literary critic John Jervis attributes the sensation to "a state of ontological undecidability or tension, where there is an insistence, a presence of whatever resists us, recalcitrant to our understanding."[52] In Diamond's case, for instance, the everyday familiarity of a maple tree was confounded by an unfamiliar sense of its subjectivity.

Freud's influential essay on the uncanny associates it with the doppelganger—the double whose presence throws one's own identity into doubt. The difficult problem of resolving one's relationship to another who is in some respects as similar as a mirror image, and yet not "oneself" as a mirror image is supposed to be, epitomized for Freud the uncanny's particular "return of the repressed." As Forbes Morlock elaborated in his essay "Doubly Uncanny," the repressed truth that returns in confronting the doppelganger is a truth about the perceiving subject: namely, that he or she is not essentially a whole, centralized, "individual" entity, but amorphous, decentralized, and with a coherence that is only contingent.[53]

The uncanny is rooted in doubts about the validity of boundaries separating self and other, or the animate from the inanimate. It evokes notions of the "living dead" or "un-dead," because it temporarily causes the perceiving subject to identify himself or herself as such. What is suddenly undead (neither living nor dead) by insistence of the uncanny double is the ideological construction of oneself that Anthony Wilden in *System and Structure* called "the Lockean ego": a self that is "autonomous in its essence" and one's personal "private property." According to psychoanalytic theory, the construction of this self is accomplished during infancy in modern Western cultures, as what Jacques Lacan called the "mirror stage": the child sees its image in the mirror and recognizes (or rather, cognizes) its "self" there, imaginatively projecting an idealized, bounded version of itself onto that image. In moments of uncanniness, the doppelganger's fusion of resemblance and nonresemblance in another object of perception temporarily destabilizes this identity by confusingly competing with our projected self-image, returning us to the repressed knowledge of its fictionality. "Intellectual certainty provides psychical shelter," Ernst Jentsch wrote in his essay "On the Psychology of the Uncanny"; but the uncanny unseats our certainty regarding the transcendent nature of our own selves and, "standing in contradiction to the usual view of psychical freedom, begins to undermine one's hasty and careless conviction of the animate state of the individual."[54] The uncanny therefore elicits at once feelings of both abjection and sublimity.

Julia Kristeva's theory of abjection characterizes it as an affect felt in the presence of aspects of the organismic, open self that were abjured in the Lacanian process of subject formation. What is abject remains proximate even while it is repeatedly but never completely estranged; Kristeva wrote, "Familiar as it might have been in an opaque and forgotten life, [it] now harries me as radically separate, loathsome. Not me. Not that. But not nothing, either. A 'something' that I do not recognize." Signaling its connections to the uncanny, the feeling of abjection typically arises when intense experiences destabilize one's identity, driving one temporarily back across the "thetic threshold" (as Kristeva renamed Lacan's mirror stage) to a more organismic and irrational state of being.[55]

In his book *Plant-Thinking,* philosopher Michael Marder cataloged the characteristics of plants that mark them as potentially uncanny. They are ubiquitously familiar, yet on second thought strange. This is perhaps particularly true of trees, whose parts we name after our own body parts: "trunk" and "limbs," "crown" and "foot." The uncanniness of plants, when they are seen to resemble us, is attributable to the sense that their subjectivity, to the extent that they can be said to have any, is predicated entirely on their being part of a larger intelligent system—a quality that we emphatically deny in reference to our own selfhood.

The potential uncanniness of plants is typically met with a psychological defense strategy of exaggerating the dissimilarity between them and us. A similar propensity was observed in 1970 by roboticist Masahiro Mori in mapping people's aversions to robotic humanoids. People preferred robots that were clearly machines, while those that were "too" human provoked a defensive revulsion that Mori dubbed the "uncanny valley."[56]

Historically, Enlightenment science and philosophy avoided the uncanny valley by exaggerating the differences separating humans from animals and plants, denying them empathy in order to minimize their uncanniness. The Cartesian Père Malebranche is said to have kicked a dog in order to demonstrate that animals were machines without feelings. Antoine Roquentin, the protagonist of Jean-Paul Sartre's novel *Nausea,* similarly reviles a chestnut tree. Roquentin reacts against the sensation of uncanniness produced in him by the tree's struggle for survival, appalled by its implications for the meaning (or lack thereof) of his own life. He vilifies the plant, calling it "monstrous and obscene," in order to alleviate his existential angst.[57]

By contrast, ecological texts focused attention on the uncanniness of plants and, by extension, on the related yet alien qualities of their subjectivity and ours. The film version of *The Secret Life of Plants* works cinematographically to subvert the viewer's habitual psychological distancing strategies, using narration, time-lapse photography, and macro-photography to keep plants uncanny. Through those techniques, it confronts its viewers with a depiction of plant sentience that intends to shake the deeply held and culturally reinforced belief that human selfhood signifies a transcendence of nature.

In *Nausea*, Sartre's hero is appalled that plants have no meaning to their existence beyond a necessity to nourish themselves and propagate. This made them existentially "absurd"; whereas a transcendent will, according to Sartre, endowed humans uniquely with the ability to make meaning. *The Secret Life of Plants* turns the tables on this thinking, implying that human sentience is but an evolutionary extension of the primal vegetal intelligence. Dr. Prem Chand of the International Plant Communications Society is quoted as saying, "*We* are the absurd appendages of an ongoing Nature, and nothing more. The plants alone prepared the Earth for all life. For what have we prepared it?"[58]

Beyond the claim to a transcendent rationality, the most time-honored basis for exaggerating the difference between human and plant subjectivity is the idea that plants do not move under their own power. This is the source of the classical distinction between "animate" and "vegetable" matter. However, plants do move; their movements of growth and decay are just not easily observable, since they occur at much slower speeds than most animal movements.[59] *The Secret Life of Plants* uses time-lapse photography to bring plant motions up to speed, making their movements perceptible, and bringing plant life more clearly into the range of the uncanny. To the accompaniment of Stevie Wonder's music, acorns sprout, pinecones swell, and mushrooms and fungi emerge from mulch. Pseudopods of slime mold dance (see plate 2), and predatory tendrils grab a nearby plant and pull it down.

In another sequence, the filmmakers invert this strategy, using time-lapse photography to suggest how the pace of modern life might appear to a vegetal sensibility. The very technique that makes plant movements perceptible to us makes our lives appear frenetically fast and loud. Highway traffic speeds by, factories belch noise and smog, and cranes at construction sites swing around madly. "Imagine receiving from plants, locked in their own dimension of time and space, a view of our own chaotic world," Prem Chand muses in the film. "To them perhaps [we are] a hopelessly mechanical rush of pointless activity . . . a flurry of the absurd."

The technique of macro-photography complements the film's time-lapse photography in its quest to keep plants uncanny. Just as time-lapse can collapse differences of pace, macro-photography can reveal hidden similarities of form and process by producing extreme close-ups that make extremely small objects and movements visible. Near the end of the film, by repeatedly cutting between extreme close-ups using macro-photography and extreme long shots using aerial photography, the filmmakers succeed in confounding the viewer's sense of scale. The differences of scale collapsed in this manner continue to grow, until a microscopic crystal might be a galaxy, and a seed photographed against a black background looks like a planet floating in space (see plate 3). Meanwhile, the narration describes the subjectivities of plant and human as two subsystems of an energy flow that passes in cyclical alternation between their two forms, as the two take turns eating each other. "They live in you, and when you die, your flesh will live again in other plants, and from those plants, more men and women fare, and on and on." Thus plant and human subjectivities are represented as different and yet related.

Music for Plants

The film *The Secret Life of Plants* reminded its viewers that, in plant-human relations, humans were the relentless predators. If we are not complete parasites in living off of plants, we nevertheless take much more from them than we give them, while we are alive. Seeing this relationship from a perspective in which plants' uncanniness is acknowledged to indicate the presence of a related subjectivity, the filmmakers hoped that we might become "more aware of *the responsibility that we have to our own*—responsibility for the food that we eat and the air that we breathe, given to us by the plants."[60] Had humans anything to give in return?

If plants were aware of their environments, it was with a quality of attention different from that of humans or animals—one that had evolved in relation to their rootedness. Plants attended to changes in ambience: qualities of the air, soil, and sunlight, as well as to other vibrations, some perhaps beyond human perception.[61] As such, music seemed to offer a possible middle ground where two such alien subjectivities as the plant and the human might meet.

The responsiveness of plants to ambient sound vibrations merited an entire chapter, "The Harmonic Life of Plants," in Tompkins and Bird's book *The Secret Life of Plants*. Beginning in the late 1950s, they wrote, numerous experimenters had reported improved growth rates in plants that were bathed in music. T. C. Singh in Tamil Nadu played Indian ragas to rice; Eugene Canby in Ontario played J. S. Bach to wheat; and George Smith in Illinois played Gershwin to corn and soybeans. Plants also seemed to flourish when exposed to sustained drones of pure tones (sine waves) at frequencies between 3,000 and 5,000 Hertz.[62] In 1976 electronic music pioneer Mort Garson recorded an album titled *Mother Earth's Plantasia: Warm Earth Music for Plants . . . and the People Who Love Them* using a Moog synthesizer.

The most famous experimenter along such lines in the seventies was Dorothy Retallack, a church organist pursuing her BA degree at Temple Buell College in Denver. In 1968 Retallack devised an experiment to test the assertion of "sound therapist" Laurel Elizabeth Keyes that the tones of B and D would revive drooping African violets. She reported in her results that those tones indeed revived violets, but they also wilted geraniums, corn plants, and philodendrons. The scope of Retallack's experiment expanded when she heard that two of her fellow students had played rock and classical music to two different sets of squash plants, and that the plants had bent away from the rock music and toward the classical. The students hypothesized that the plants grew toward what they liked and away from what they didn't. In the summer of '69, Retallack followed up on that experiment, pitting hard rock against easy listening and measuring their effects on petunias, zinnias, marigolds, corn, and squash. She reported that after two weeks the plants in the easy listening chamber were "lush," with a thick root network, and leaning toward the radio. Those subjected to hard rock, by contrast, were "confused" and bent into "grotesque shapes," with yellowing leaves and poor root structure. After another week they died. "Some unknown force had crippled, then destroyed these plants," she concluded.[63]

After the college's publicity office offered Retallack's story to the *Denver Post* in 1970, her experiments gripped the popular imagination. The resulting *Post* story was syndicated and reappeared over the following year in newspapers around the country. The *CBS Evening News* convinced Retallack to repeat her experiment so that they could film it using time-lapse photography. The results were broadcast in October 1970, on a news day in which viewers also learned that a grand jury had absolved the National Guardsmen involved in the Kent State shootings; that martial law had been declared in Canada to suppress the Quebec Liberation Front; that the judge in the My Lai Massacre trial had refused to admit testimony from witnesses who testified before Congress; and that in response to bombings and terrorist threats, the federal government would start inspecting all packages and briefcases brought into federal buildings. Finally, Walter Cronkite reported, Retallack's experiments indicated that "rock music, like smoking, may be dangerous to your health."[64]

However, in her 1973 book, *The Sound of Music and Plants*, Retallack emphasized the positive effects that music could have on plants. She wrote that they responded positively to Johann Sebastian Bach's preludes and to jazz. They fared the worst when subjected to white noise and were neutral toward twelve-tone avant-garde music by Arnold Schoenberg, Anton Webern, and Alban Berg. Above all they responded best to Hindustani ragas played on the sitar by Ravi Shankar.[65]

Retallack identified two common factors in the styles of music that her plants seemed to like best: they were improvisational, and they were sacred. In this connection she quoted Shankar's explanation of the meaning of raga in his 1968 book, *My Music, My Life*: "The highest aim of our music is to reveal the essence of the universe. . . . Through music, one can reach God." Retallack added to this, "Music and plants and people can use bridges between each other or *be* the bridges themselves for greater harmony. There can be a bridge of science to music; man to nature; man to himself; man to man; nation to nation; and man to the Infinite." Religion and science, she continued, were once unified pursuits, but then they "drifted apart and degenerated." Music might now constitute a bridge to "a new Religion-Art-Science . . . which can find what has been lost." On this point, Retallack's belief system intersected—somewhat surprisingly—with the visions of Terence and Dennis McKenna, who believed that by tuning in to the ecosystem through psychedelics and a musical "chant-induced, collective synesthesia," humanity could discover (or rediscover) a holistic, post-Enlightenment "Orphic science" akin to shamanism.[66]

Retallack, citing Cleve Backster's work on "primal perception," contended at the conclusion of her study that plants had been demonstrated to be both sentient and communicative, and it was likely that they had some wisdom to impart. "The plants, it seems to me, are trying to tell us something," she told the *CBS Evening News*. "Music was the subject; plants were the bridge—or was it the other way around?"[67]

FIVE

Ambient Music

A jet scrapes the sky over my head and I ask: "Yes, but is it music? . . ."

R. MURRAY SCHAFER, *THE NEW SOUNDSCAPE*

Ecological thinkers of the seventies understood nature as an intelligent, sentient system. In conjunction with this, environmentalism asserted that the future survival of our species depended on our cultivating better relations with this intelligence. This meant learning how to live in harmony with the multiple, complex, and subtle feedback loops that sustained the system's health and, in the process, our own. But how were modern Americans to acquire this necessary learning? Our science was crippled by a congenital rejection of nature's sentient qualities. The best way therefore seemed to be by engaging in an intersubjective dialogue with sentient nature itself.

But how to conduct this dialogue? Through what medium of communication? How were Americans to orient themselves to be able to listen properly? And how could they voice their own needs without violating nature's integrity? Some ecological thinkers advocated adopting or adapting Native American ways of living. Others tried communing with plants. The composers of ambient music developed a third approach.

Using ambient sound as a means of dialogue with nature's systemic intelligence involved integrating two kinds of practice. One was listening. Many believed that humanity in general would have to quiet down, so that nature could once again hear itself think. The other was voicing. This involved acknowledging that, when nature thinks, humans are also part of that thinking process. How could human society find its voice *in* nature? Experimental composers invented various ways of making music that interfaced with nature's systemic intelligence by emulating its feedback processes. They used feedback loops to integrate the sounds of daily living into a living environment of sounds.

Sonic Meditations

In 1971 Dennis McKenna, high on psychedelic mushrooms in the Colombian rain forest, heard a buzzing in his head. It was "on the absolute edge of audible perception"—barely even a sound at all. But perhaps, he thought, it was a communication. So, as he recounted, "I tried to imitate these noises with my vocal chords, just experimenting with a kind of humming [and] . . . [s]uddenly it was as if the sound and my voice locked onto each other and the sound was my voice . . . [now] suddenly much intensified in energy." This experience of discovering a resonant frequency that matched the sound in his head led McKenna to theorize that certain sounds of the human voice could constitute a link or "interphase between consciousness active in the world [the intelligence of nature], and consciousness active in the central nervous system [the human individual]." This was because, in communications between the individual and the ecosystem, "the intermediary is the body."[1]

In this anecdote, McKenna describes responding to an ambient sound at the threshold of audibility by meeting it with his own similar tone. While he may have discovered it on his own, as a musical practice of give-and-take with the environment, the same technique had been pioneered by composer Pauline Oliveros during the 1960s. Oliveros wrote several pieces organized around feedback loops of listening and voicing at the threshold of audibility, which she called "sonic meditations." In 1974 she published the scores of several such works. Among them is *Sonic Meditation VIII*, which directs the performer to use a single-pitch drone (as McKenna did) to establish an "environmental dialogue": "As you become aware of sounds from the environment, gradually begin to reinforce the pitch of the sound source or its resonance. If you become louder than the source, diminuendo until you can hear it again."[2]

Oliveros called her pieces "meditations" in order to distance herself from aesthetic concerns in favor of spiritual ones. Music making, she wrote, should be inclusive, participatory, and healing, rather than exclusive and artistic. She described the motivation of her musical practice as "communication among all forms of life through Sonic Energy." The process, she explained, entailed giving up intentional control over the outcome of the practice and what it sounded like: "In process a kind of music occurs naturally. Its beauty is not through intention, but is intrinsically the effectiveness of its healing power." This beauty would perhaps not be perceived by an outside listener, but would be felt by those participating in the process.[3]

Oliveros was one of a group of composers in the sixties and seventies who explored the uses of music as a decentralized physical phenomenon, which derived its meaningfulness on that basis rather than, as was traditional, from an intentional arrangement of sounds. For them, music was a way of integrating sounds into a natural, open system, organized by feedback relations rather than having been put in order by the composer's dictatorial authority. In this way the music would take a form that integrated it with other natural, open systems, and it would be, Oliveros wrote, "a language which can fit the fluid process" of nature.[4] Through music that was itself an open system, humanity could meet and interface with the systemic intelligence of nature.

Historically, Oliveros, La Monte Young, and Terry Riley were perhaps the most important pioneers of this musical form. Other important contributors to its evolution were Max Neuhaus, Alvin Lucier, Cornelius Cardew, Richard Teitelbaum, David Rosenboom, Stuart Dempster, Steve Reich, Michael Nyman, Nicolas Collins, and Charlemagne Palestine. In the second half of the seventies, Brian Eno and Philip Glass were instrumental in bringing it to the attention of the general public.

As a practice, this style of experimental composition never strayed far from the social and environmental concerns that characterized other manifestations of the culture of feedback. It was Glass who composed the soundtrack for Godfrey Reggio's film *Koyaanisqatsi*. According to musicologist Mitchell Morris, the "inexacted repetition" of Glass's score reinforces the cinematography in presenting the natural and human worlds as two systems characterized by different dynamics. Morris wrote that the music's evolving and variegated "flows" and "accretions" offer "just enough regularity to feel like complex instantiations of underlying order."[5]

Music history has been slow to distinguish the body of work associated with ecological thinking from other musical experiments of the period. Among musicologists, the term "minimalist" has come to delineate a style of composition with which this body of work often intersects; but as Glass asserted in 1980, the notion of musical minimalism is a misrepresentation when applied to what is in fact a "music that's based on process" and "repetitive structures." More aptly, Joan La Barbara in 1974 grouped Young, Riley, Reich, and Glass together as "the Steady State School." Tom Johnson in the *Village Voice* in 1972 referred to the same four composers as the "hypnotic" school, "because it is highly repetitious, and employs a consistent texture, rather than building or developing in traditional ways." Johnson continued, "Usually pieces in this genre are rather long, and they can seem tedious until one learns how to tune into the many subtle variations which go on underneath the sameness of the surface. Then very new and exciting musical experiences begin to happen."[6] In popular culture, "ambient" or "ambient drone" is a name commonly given to this music, which I also adopt here.

In order to dispel the confusion engendered by the use of the terms "minimalism" and "avant-garde" in reference to this musical form, Michael Nyman in his 1974 study and "eyewitness account" of the music, *Experimental Music: Cage and Beyond*, uses the term "experimental," which I also prefer. In his book's first chapter, titled "Towards (a Definition of) Experimental Music," Nyman offered examples of the various processes by which composers of this school constructed open systems in order to embrace the indeterminacy advocated by John Cage while moving beyond Cage's use of chance operations.[7] These strategies included modular repetition, the use of contextual cues, and electronic feedback loops.

Roger Johnson followed Nyman's lead in his 1981 anthology of new music and also designated this music "experimental," as distinct from the "avant-garde" tradition centered around Arnold Schoenberg, Anton Webern, Karlheinz Stockhausen, Pierre Boulez, and Milton Babbitt. The latter school was predominantly European, he wrote; the former, more American. According to Johnson, work in the experimental vein was defined by (1) open forms or indeterminacy "exploring different blends of structure and spontaneity"; (2) an expanded sound vocabulary to include new (electronic) or previously unnoticed or unmusical sounds; (3) new environments for musical performance; (4) "meditative and spiritual practices and expanded states of consciousness" explored through listening and voicing; and (5) "gradually unfolding compositional processes, patterns, and phase relationships."[8]

In my opinion, the most important characteristics of this body of music are these two: (1) regarding music as a physical activity and listening as a psychophysiological phenomenon; and (2) using reiterative processes to produce emergent, cumulative sonorities in which chance events coexist with multiple overlapping chains of causality. These principles account for the key formal features of a variety of compositions. A widely shared determination to explore the "threshold of audibility" was motivated by interest in the psychophysiological dynamics of interaction between a listener and a place. Reiteration emphasized the role of feedback in creating an emergent order. Mutating musical reiterations at the threshold of audibility could integrate ambient noises into an open system of sounds, creating a sonic field in which human sounds are embedded.

Noise as Pollution

The project of interacting with natural systems through sound was made problematic by the presence of high levels of ambient noise. In the early seventies, noise was identified as a form of pollution, the equivalent of more tangible pollutants like litter and chemicals: a negative feedback generated by the excesses of industrial civilization. In *The Book of Noise*, published in 1970, Canadian composer R. Murray Schafer compared noise to human waste, calling it "sonic sewage."[9] Noise was officially recognized as a form of pollution in the United States in 1972, with the passage of the federal Noise Pollution and Abatement Act. As a result, the Environmental Protection Agency included an Office of Noise Abatement and Control, which launched advertising campaigns to raise public awareness of the problem.

Jet engine noise proved a focal point for environmentalists' struggle against ambient noise. The heavens, Schafer complained, once a symbol of peacefulness and freedom, were now oceans of unwanted and uncontrollable sound waves—victims of the tragedy of the commons. "Any number of thunderous things can happen in the sky over our heads without restriction as to how frequently or how loudly they may happen," he complained. "The whole world is an airport. What are we going to do about it?"[10]

In particular, the environmental movement's successful fight to end federal funding for the Supersonic Transport jet—the SST—galvanized noise abatement activism. For the first Earth Day in 1970, Laurence (Larry) Moss, who was on the board of directors of the Sierra Club, organized the Coalition Against the SST, which brought together legislators, environmentalists, and scientists. Although cost questions loomed large in the minds of legislators, public discourse centered on the noise issue. Matthias Lukens, representing the operators of the country's major airports, stated that SST funding should be withheld until engineers resolved the "key question of noise" during takeoff and landing. In congressional hearings, Moss emphasized the "sideline noise" of the SST's engines, noting that it would be "highly objectionable at a distance of over 15 miles." He prepared maps, which were distributed to the public by the Federation of American Scientists, showing how the metropolitan areas of New York, San Francisco, Seattle, Honolulu, Anchorage, Boston, and Los Angeles would be affected.[11]

Even more damning in the public imagination than the SST's engine noise was the effect known as the "sonic boom." A plane flying at supersonic speed pushes a cone of compressed air in front of it that topples like a sound-tsunami onto everything nearby. According to government-funded studies, "all available information indicates that the effects of the sonic boom are such as to be considered intolerable by a very high percentage of the people affected." In March 1967, two Harvard scientists, physicist William Shurcliff and biologist John Edsall, founded the Citizens League Against the Sonic Boom (CLASB). For the struggle against the SST, CLASB bought ads in the *New York Times*, the *Washington Post*, and the *Wall Street Journal*. It also published the *SST and the Sonic Boom Handbook*, which asserted the likelihood that American neighborhoods and territorial waters would soon become "a vast dumping-ground for sonic booms."[12]

Public relations campaigns concerning noise pollution tended to emphasize its effects on individuals. This was seen as the best means to generate widespread support for noise-abatement regulations. The perspective cultivated by this strategy might be summarized as "Quiet down, and let me hear myself think." As a conceptualization of the relationship of the individual to noise, however, it reinforced a mentality that Vine Deloria Jr. disparagingly referred to as "the isolated individual helplessly gripped by supraindividual forces."[13] What was missing was the sense that everyone who suffered from the noise of industrial civilization was also benefiting from and contributing to it.

An animated public service announcement made for television in 1977 exemplifies this attitude. In it, a commuter dressed in a suit and tie and carrying a briefcase is subjected to a barrage of traffic noises on his way to work. The squeal of the subway train on its tracks causes him to grit his teeth and squeeze his eyes shut in agony. Once he emerges onto the sidewalk, the noise of street traffic—passing motorcycle and car engines, squealing tires, and honking horns—slices through his skull. Significantly, however, all these sources of noise are invisible.

Even the subway and the street are vacant except for the ad's lone protagonist. Although he is in a public space and using public transportation, all of the other people using them too are not visually represented. They are indicated only by the noise they create (see plate 4). The effect is to create the impression that rather than being a member of a loud civilization, the protagonist is a lone individual persecuted by a mass of noisy others. Likewise, in the final scene, the protagonist is enjoying his leisure alone in the woods, photographing a bird. The noise of a chain saw interrupts, causing the bird to fly off and splitting the man in half, like a sawn tree, down the middle. This equates him visually with a transgressed-against nature, rather than with the human sources of noise; his two halves fall down as we hear the sound of the sawn tree creaking and then crashing to the ground. In the ad's final image, his two toppled halves reconverge, resurrected by "quiet"; and he stands looking out at us imploringly, his finger to his lips in a plea for silence. The harm done to him by noise is presented as a version of the Romantic opposition between the individual and society, rather than as a problem caused by a society of many individuals just like him.

Similarly, an EPA activities booklet distributed to schoolchildren in 1980 featured a "Noise Maze" that challenged the reader to find "the most quiet" path to school by avoiding noise—which, the text emphasized, leads to "STRESS, TENSION, and ANXIETY."[14] The wrong paths go past a jackhammer, a factory, a subway, construction, an airplane, a highway, and yelling people. The correct path goes past trees, swings, a lake, a golf course, and a church (fig. 8). The maze in this way creates structural oppositions pitting nature, spirituality, and leisure against industrial work and transportation. In both the televised PSA and the activities booklet, noise pollution was defined as the sounds of industrial civilization that intrude on and upset one's enjoyment of the natural soundscape. Significantly, in the animated television ad, the birdsong that is heard while the protagonist is clipping roses does not figure as noise, although it is louder than the noise of the barking dog that does.

A related argument for noise abatement that was present to a lesser degree in the PSA may be summarized as "Quiet down and let nature hear itself think." In a small but salient detail, the bird that flees from the noise of the chain saw in the final scene returns to perch on the protagonist's shoulder once quiet is restored (see plate 5). By ending noise pollution, the image suggests, humanity can join with nature instead of living "against" it.

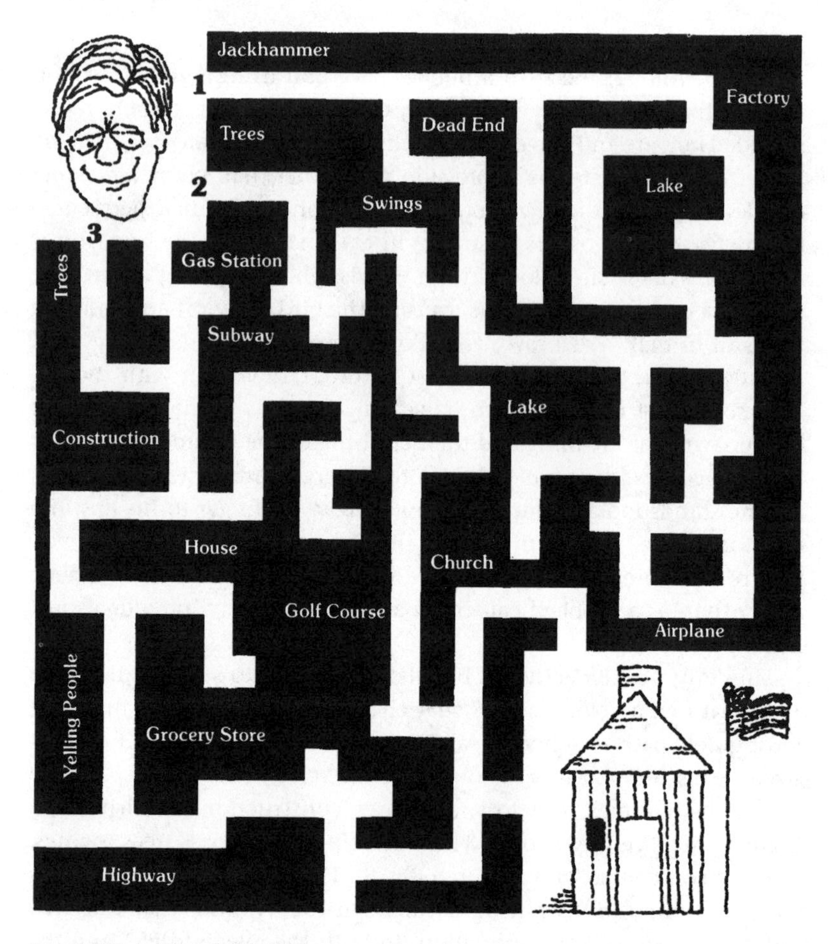

Figure 8. "Noise Maze," EPA, 1980.

This was the outlook of R. Murray Schafer and his followers in the "acoustic ecology" movement. Schafer advocated mitigating noise in order to "let nature speak for itself. Water, wind, birds, wood and stone." Motor noises had become almost omnipresent, Schafer wrote, and the effect was to obscure natural sounds—especially the sound of moving water like rivers and oceans. The sound of moving water had dominated the aural environment in pre-technological eras, he believed, and it had offered humans a subliminal reminder of natural cycles. By contrast, motor noises symbolized a warped idea of progress that equated it with technological advances enabling the ever more rapid exploitation of natural resources. "Every society of the past has had some built in philosophy of restraint," wrote Schafer, but contemporary American culture celebrated progress as freedom from natural restraints; hence our "noise is a fitting symbol for a vulgar epoch." Americans now faced a cultural challenge, he believed, to redefine progress as advancing toward a lifestyle of less exploitation and effluence. A "more emphatic public interest in the sonic environment" would be an essential factor in bringing about that cultural change.[15]

Acoustic Ecology and "Schizophonia"

Schafer understood noise production to be rooted in a destructive feedback loop: noise was both the cause and the result of not listening carefully to our aural surroundings. Unfortunately, he wrote, the most common strategy for coping with noise was to block it out; and we have become so adept at unconsciously filtering out noises that we have become "careless" of the soundscape. For example, when the residents of Vancouver, British Columbia, were asked in 1969 to estimate the number of times seaplanes flew over their homes in a day, the average guess was eight per day; the actual count was sixty-five. As Schafer wrote in his book *The New Soundscape*, "noise is sound we have been trained to ignore."[16]

Unconsciously filtering out unwanted sounds and listening to remote entertainments that blocked ambient noise (car radios, for instance) dissociated people from their immediate environments. Schafer called this "schizophonia." In particular, he singled out for criticism the use of music as "*audioanalgesia*, that is, the use of sound as a painkiller, a distraction of the senses from the real facts of existence." This was a form of schizophonia epitomized by Muzak, which piped innocuous music into potentially anxiety-producing spaces like elevators, airports, and department stores. Schizophonia offered short-term benefits but at long-term costs. The environmentally winning strategy, Schafer argued, was for people to open themselves fully to the environmental soundscape and take responsibility for shaping it. As a practice opposed to that of schizophonia, he advocated "ear cleaning: a systematic program for training the ears to listen more discriminatingly to sounds, particularly those of the environment."[17] He propounded this approach in two books published in the late 1960s, *Ear Cleaning* (1967) and *The New Soundscape* (1969).

Ear cleaning, Schafer asserted, would reinsert noise pollution into a constructive feedback loop. The improved hearing produced by ear cleaning would foster a conscious crafting of the soundscape, as people began to choose consciously which sounds they were willing to live with. "The boring or destructive sounds will be conspicuous enough," Schafer wrote, "and we will know why we must eliminate them"; listeners would also choose "which sounds . . . we want to preserve, encourage, multiply." Schafer called this conscious shaping of the soundscape "acoustic design."[18]

Like much of the environmentalist movement, Schafer's acoustic design project harbored a built-in anti-industrial and anti-urban bias. The cleaned ear, he believed, would lead people to choose soundscapes of "elegance and beauty"—aesthetic qualities that, in his mind, primarily characterized natural sounds as opposed to industrial ones. In his 1977 work, *The Tuning of the World*, Schafer criticized the urban soundscape in a way that disparaged its inhabitants, referring to "the slop and spawn of the megalopolis" and their "multiplication of sonic jabberware." Against them he positioned the acoustic designer, charged with the task of "sorting out the mess." Acoustic designers took the pre-industrial rural soundscape as their ideal. They assumed that such soundscapes represented the proper balance of human and natural sounds.[19]

This bias was evident in the innovative social use that Schafer envisioned for broadcast radio. Commercial radio broadcasting accelerated people's lives and consumption patterns with its pop, patter, and hype, he wrote. Instead, radio could be repurposed toward "reinforcing once again the natural rhythms of life"; it "could be employed to show us alternative modes of living." This would be accomplished by transmitting natural and rural sounds into urban spaces. Acoustic designer Bruce Davis, one of Schafer's followers and a participant in Schafer's World Soundscape Project, called this "Environmental Radio." Between 1972 and 1975, the World Soundscape Project began to realize this vision by embarking on a mission to document and preserve various "endangered" rural soundscapes through sound recordings that could later be broadcast. A tour across Canada in 1973 resulted in a series of ten hour-long broadcasts over CBC radio in 1974, billed as *Soundscapes of Canada*. This was followed in 1975 by a tour of Europe in which Schafer and his colleagues recorded rural soundscapes in Germany, Italy, France, Sweden, and Scotland.[20]

However, if one takes seriously the idea that the soundscape is nature's means of hearing itself think, the World Soundscape Project incorporates a troubling bifurcation of the human subject in its engagement with this process. In the acoustic designer's ideal world, the "listening" human subject, equipped with microphones and recorders, is part of a technologically advanced civilization; but the "voicing" human subject, at one with the bucolic surroundings, is insistently pre-modern and pre-industrial. This suggests a kind of schizophonia analogous to multiple personality disorder. The resulting "dialogue" between humans and nature has the quality of a museum archive, not a living relationship.

Acoustic designers approached the problem of acoustic ecology with a mind-set that Margaret Mead and Gregory Bateson had denounced as "social engineering." That is, the acoustic designers assumed that they knew ahead of time what the end result of ear cleaning should sound like. What if one were instead to undertake ear cleaning while following Bateson's dictum to "discard purpose in order to achieve our purpose,"[21] engaging in a process of deutero-learning without dictating the outcome?

Pursuing an approach that diverged in this respect from that of the acoustic designers, experimental composers like Pauline Oliveros thought about sound as a phenomenon of nested systems in which humans are holons. They let the property of macrodetermination that is characteristic of nested systems decide what the results of their interventions in the soundscape would be. This strategy allowed them to avoid the engineering mind-set and the bifurcation of the voicing and listening human subject that were implicit to acoustic ecology. They asked, how can we bring nature to hear itself think, when we include ourselves as a part of that nature? How can one hear oneself as part of that thinking process, and play one's role in it without monopolizing the conversation?

Oliveros approached the solution to this problem through a process that entailed the possibility of simultaneously listening and voicing, bringing herself in the present moment to play a part in the soundscape. The score of her piece *Sonic Meditations XIII* from 1971, titled "Energy Changes," instructs the participant to become aware of every sound in the room, both external and internal; then, "when there's a feeling for making a sound," she produces one. Afterward she immediately begins to restore the previous quality of listening: "When the sound is produced, then they have to return and reconnect with hearing everything that they were hearing before they made the sound."[22]

The ability to alternate instantaneously between listening and voicing depended on crossing back and forth between the two at the threshold of audibility. As Oliveros wrote in "The Poetics of Environmental Sound," first published in 1970, "You are part of the environment. Explore the limits of audibility." Significantly, Schafer's ear-cleaning exercises, like Oliveros's "sonic meditations," also involved listening at the threshold of audibility. The difference lay in that Oliveros's meditations always included voicing there, as well. As she wrote in 1972, "Transmit at the threshold. Feed back. . . . You must become the receiver to transmit at the threshold of audibility. You must become the transmitter in order to receive at the threshold of audibility."[23]

Ambient Sound

The possibility of integrating human activity into the soundscape by cultivating a quality of attention that could comprehend simultaneous listening and voicing allowed experimental composers to problematize the meaning of noise. What is the sound that should be eliminated? Thinking about what is meant by "noise" instead of assuming that it is an objective category to which certain sounds (like the buzzing of chain saws) intrinsically belong foregrounds issues of psychoacoustics, or the subjective aspects of hearing. There is an element of subjectivity involved in assessing the noisiness of noises.[24] This extends to the question of whether or not a sound is even a noise at all. "Noise" might be an important source of information, or an unintended addition to the diversity of aural experience.

In 1974 experimental composer Max Neuhaus penned an open letter in response to a pamphlet titled "Noise Makes You Sick" promulgated by New York State's Department of Air Resources. Neuhaus's response was printed in the *New York Times* editorials section. In it he argued:

> No sound is intrinsically bad. How we hear it depends a great deal on how we have been conditioned to hear it. Through extreme exaggeration of the effects of sound on the human mind and body this [anti-noise] propaganda has so frightened people that *it has created "noise" in many places where there was none before*; and in effect *robbed us of the ability to listen to our environment.*[25]

For over a decade beginning in 1966, Neuhaus had conducted walks through New York neighborhoods, billed as musical performances. A 1976 poster advertising one such walk, captioned with the simple directive "Listen" (fig. 9), features the Brooklyn Bridge photographed from an intriguing but unlovely perspective. Instead of the familiar panoramic view of the bridge's suspension towers, we are treated to a view of its underside, with crisscrossing roadways and walkways, and

Figure 9. Max Neuhaus, *Listen*, 1976.

tall buildings in a diversity of architectural styles framing its black mass in a web of complexity. Neuhaus explained, "It came from a long fascination of mine with sounds of traffic moving across that bridge—the rich sound texture formed from hundreds of tires rolling over the open grating of the roadbed, each with a different speed and tread."[26] The bridge in this poster, like the city beyond it, is looming and loud, yet not malignant.

Like Arne Næss's "deep ecology," Neuhaus's approach to soundscapes valued complexity and diversity.[27] In lieu of whitewashing the urban soundscape with the sounds of rivers or village church bells, as acoustic designers hoped to do, Neuhaus approached it as a complexity to which feedback loops could bring an emergent order. For him, as for other experimental composers, the issue was not whether (or under what conditions) a sound could be annoying, but how one could use sounds to orient oneself (psychologically and politically as well as directionally) in relation to one's social and physical environment.

To this end, experimental composer Alvin Lucier adopted the metaphor of echolocation, a biological sonar that some animals use. His *Still and Moving Lines of Silence in Families of Hyperbolas* from 1974 required performers to use sound waves to navigate the performance space. This was presented as a learning process that Lucier intended also to have a social dimension. As he wrote:

> I would eventually like to make realizations in which many players, singers, and dancers perform at the same time, so that every player would be in as complex a situation as that of the water-skimmer [which produces wavelets to navigate a pond]. If the water-skimmer is alone in a pond, it's in a very simple condition, but the minute you add another skimmer, the first one has to perceive echoes from the edge of the pond, with all that that entails, plus echoes from the second water-skimmer. That's the situation, the natural situation, that I want the piece to achieve.[28]

Lucier's *Gentle Fire* of 1971 entailed the transformation of a variety of everyday "ambient sound events" into other ones by electronic means. Lucier later said:

> The first column has images that are supposed to be unpleasant, and those in the second are supposed to be pleasant, but I can't decide whether some are pleasant or unpleasant, so I put them in both. If you paired them together, I don't know how you'd deal with them. Perhaps you could just change your mind about how you felt about them and the exchange could be made mentally.

In describing the relationship between *Gentle Fire* and the problem of noise pollution, he told an interviewer:

> I feel guilty about *Gentle Fire* because one of the ideas is that you can learn to tolerate noise and pollution . . . but why wouldn't I prefer to take political action to stop noise pollution instead of allowing it to happen, merely dealing with it in a dreamy way? . . . This is difficult. It's got something to do with the way things are changing. . . . [I]nstead of . . . the idea of contrast or competition, we're shifting to ideas of simultaneity, or similar identity.[29]

Neuhaus's and Lucier's approach to noise and music followed a path blazed by John Cage, the most influential experimental composer in mid-century America. Cage was not interested in using composing to direct the listener's attention by the manipulation of an aural "message." He aimed instead to change the listener's idea of what listening could be. His works offered listeners lessons in deutero-listening.

The Western musical heritage, Cage asserted in his 1961 manifesto *Silence*, represented deeply ingrained listening habits. Cage wanted to challenge his listeners to discard their habitual expectations and value judgments concerning sounds and instead to adopt a frame of mind open to all experience. He called for cultivating "a purposeful purposelessness" in composing and listening to music, corresponding to a quality of attention connoting "an affirmation of life." Such an affirmation entailed accepting the diversity of all sounds, rather than making judgments regarding their desirability. His compositions therefore allow no distinction to be made between "musical" sounds and "noise." He wanted his works to lead the listener, by way of this new quality of attention, "to the world of nature, where gradually or suddenly, one sees that humanity and nature, not separate, are in this world together." If the change in listening habits that Cage promoted were to take hold, the listener would always be surrounded by music; but what was meant by music would be radically redefined in the process.[30]

Cage was also revolutionary in decentralizing the practices of musical composition and performance. He worked to liberate composing, performing, and listening from the control of a centralized authority (whether of the composer or the conductor), by introducing elements of indeterminacy into those processes. By "giv[ing] up the desire to control sound," he wrote, we "let sounds be themselves."[31]

Experimental composers influenced by Cage and by ecological thinking created compositions that were intended to draw the entire soundscape together through feedback loops, integrating diversity and indeterminacy through systems of sound. Max Neuhaus "composed" by building installations that created modified aural environments for passersby who had not intentionally sought out a musical experience. In 1973, for instance, he created *Walkthrough* at a subway entrance in Brooklyn, which emitted "a concert of clicks" to all passersby. The clicks were generated by a computer that determined its rate of clicking by monitoring aspects of the natural environment outside of the subway entrance, including wind speed, light intensity, temperature, and humidity. Neuhaus was pleased with the way that this piece integrated inputs from the natural environment to modify the existing soundscape in small ways, subliminally informing the people who used the subway every day about the natural world outside their tunnel. He saw it as a template for a possible kind of intervention in the soundscape, "working with a fine level of subtlety . . . [involving] very small changes in a very familiar environment, one we encounter daily."[32]

Richard Teitelbaum also undertook to enhance listeners' attention to the sounds of their immediate environment, with performances that were similar to Murray Schafer's "ear cleaning" but that embraced the diversity of the urban soundscape. His *Threshold Music* from the mid-1970s, like Pauline Oliveros's *Sonic Meditations XIII*, supplemented ambient sounds with congruent ones played at similar volumes, so that the act of listening to his musical performance directed one's attention to all the environmental sounds already hovering at the psychoacoustic threshold of audibility. As Teitelbaum explained, his score required performers "to match the level of sound of the environment as precisely as they can, as closely as they can." Tom Johnson reported in the *Village Voice* on how this worked in one of Teitelbaum's solo performances in 1975:

> Richard Teitelbaum began his May 20 concert at 224 Center Street without making any sound at all. He took his place at his souped-up Moog synthesizer, started a tape recorder, and turned a few dials, but nothing happened. For a while I thought he was having trouble with his equipment, but then I happened to notice a faint humming and gradually realized that this almost inaudible sound must be Teitelbaum's music. I began listening harder, straining to hear other musical elements. In the process, of course, I began to hear a lot of other sounds I hadn't noticed before. A couple of floors above, some machines were running, stopping and starting at odd intervals. Somewhere a long ways away a trumpet player was practicing. Occasionally a passing truck became a major sound event.[33]

Ambient Drone and Just Intonation: The Influence of Indian Classical Music

The experimental project of dynamically listening and voicing at the threshold of audibility relied heavily on the use of humming or droning sounds that were sustained for indefinitely long periods. The drone was understood to facilitate moving between voicing and listening and the two different qualities of attention they required. Oliveros described her sonic meditations as originating in 1970 out of an experience of moving back and forth between those two qualities of attention—one narrowly focused and one diffuse and open-ended—while sitting beside a public fountain: "There were sounds that crept up on me, coming out of the drone, sharing the stage with or stealing it from the fountain, and then blending themselves unnoticed back into the drone."[34]

The drone constitutes an aural "background" into which all sounds that the attention is not focused on become integrated. It is a sound corresponding to long durations: a sonic blend of the rhythms of nature that may at first seem static to the human listener. It also marks the "thetic threshold" (as Julia Kristeva named the assumption of identity that is prerequisite to an act of voicing) of human intervention in that sonic field. La Monte Young, a dedicated proponent of ambient drone music, stated in 1964, "In the life of the Tortoise [on whose back the world rests in Hindu cosmology] the drone is first sound. It lasts forever and cannot have begun but it is taken up again from time to time until it last forever as continuous sound."[35]

Young's quest for "higher aural awareness through sustained harmonies and/or drones" began, like that of Oliveros, in Northern California, where he was her classmate in the graduate program in musical composition at Berkeley in the late 1950s. Influenced by the San Francisco Beats, Young and another classmate of theirs, Terry Riley, mixed musical composition with psychedelic drugs in a quest for enlightenment. They smoked marijuana and ingested peyote together, moving on in the sixties to psychedelic mushrooms and LSD. Peyote, Riley later told historian Keith Potter, first enabled him to see "the sacredness of music." Young similarly claimed that through psychedelics he "could see that *sounds and all other things in the world were just as important as human beings* and that if we could to some degree give ourselves up to them . . . we can experience another world."[36]

Young's interest in the micro-acoustics of sustained drones made him a proselytizer of "just intonation" tuning, which is a way of tuning musical instruments that conforms to the natural relations among sound waves. In just intonation, every interval between two pitches corresponds to a whole number ratio between their wave frequencies. For example, doubling or halving the frequency produces the interval of an octave; a ratio of 3:2 produces the fifth ("perfect" fifth); and a ratio of 4:3 produces the "perfect" fourth. Because of their whole-number ratios, these intervals produce harmonic overtones that resonate.

In just intonation, the intervals between the twelve notes of the chromatic scale are not all equal; therefore each instrument in just intonation is only in tune for one key. Beginning in the eighteenth century, however, harmonic progression became an important structural principle in European music, requiring many modulations from key to key within a single piece. Therefore European musicians adopted tuning modifications known as "temperament." Tempered instruments are tuned to be slightly out of just intonation, so that in modulating from key to key, no extreme dissonance is encountered. Johann Sebastian Bach's *Well-Tempered Clavier* promoted tempered tuning by celebrating its versatility.[37] However, much of the natural resonance of just intonation is sacrificed in tempered tuning.

To Young's way of thinking, musical temperament was a perversion of our relations with nature through sound. Just intonation, based on the harmonics produced naturally by vibrating objects, better fit the needs of composers attempting to develop sound dialogues with natural systems. Young's answer to Bach's *Well-Tempered Clavier*, which he titled *The Well-Tuned Piano* (1964), made extensive use of natural resonances. In what Young referred to as the "cloud" sections of the piece, he created a feedback system by hitting the piano keys in tandem with the acoustical beats that emerged from those resonances. This resulted in "undulating clouds of sound emanating from somewhere other than the struck strings themselves." Terry Riley's embrace of just intonation dated from the fall of 1965, when he moved to New York City and renewed his friendship and musical collaboration with Young. He used it in his most significant works of the seventies, including *Persian Surgery Dervishes* (recorded in 1971–72), *Descending Moonshine Dervishes* (1975), and *Shri Camel* (1976).[38]

Experimental composers' interest in natural harmonics and just intonation was linked to their explorations of world music. Tempered tuning, although ubiquitous in European music since the eighteenth century, was not adopted elsewhere around the globe. Stuart Dempster, who along with Oliveros and Riley was a member of the San Francisco Tape Music Center in the sixties, introduced the didgeridoo of the Australian aborigines to American audiences. His 1976 compositions *Standing Waves* and *Didjeridervish* consist wholly of sustained drones supplemented by resonating natural harmonics (perfect fourths and fifths). Hindustani classical music, though, was the most widespread source of inspiration for composers looking beyond the Western musical tradition. La Monte Young heard Ustad Ali Akbar Khan's *Music of India: Morning and Evening Ragas* on the radio in Los Angeles in 1957 and immediately went out and bought it. The drone of the tambura, the typical accompanying instrument in Hindustani music, inspired

Young's subsequent sustained-note compositions, beginning with 1958's *Trio for Strings*.[39]

In 1970 Young began the study of Hindustani music in earnest by becoming a disciple of Pandit Pran Nath. Among the various schools of Hindustani classical music, the Kirana gharānā to which Nath belonged focuses particularly on mastering the precise tuning and emotional expression of individual tones. Young first heard Nath's singing in 1967, in a recording given to him by his yoga teacher. In January 1970, he and his wife, Marian Zazeela, arranged for Nath to visit New York City and perform. Both of them became Nath's disciples at that time. Next they brought him to California to visit Terry Riley, who then became Nath's disciple as well. In September 1970, Riley moved with his wife and child to India to spend six months living with Nath, studying raga rhythms and vocal intonations. Young and Zazeela followed him three months later. In the autumn of 1971, Riley began teaching at Mills College in Oakland, California, and he started a program in Indian music there at which Nath lectured periodically. Between 1971 and 1977, Young, Zazeela, Riley, and Nath regularly performed Hindustani music together for audiences, with Nath singing, Riley on the tabla, and Young and Zazeela playing tamburas.[40]

The culture of Hindustani classical music emphasizes its ties with spirituality and natural cycles. According to Hazrat Inayat Khan, through the raga both performer and listener "become absorbed in the whole and single immanence of nature." Like all Indian classical music, its performance is rooted in Hinduism's Vedic chants. But Sufism is also a significant influence in the Hindustani raga form particularly, and especially in the style of the Kirana gharānā. Hence Riley's references to dervishes in his compositions *Persian Surgery Dervishes* and *Descending Moonshine Dervishes*, alluding to the Sufi practice of holy men seeking unity with god through ecstatic trance states. Riley later clarified that although classical Indian music was not the only source of his mature style, it helped him to grasp the "integrated patterns and the way they move against each other" that he had been attempting in his compositions.[41]

The raga could become a tool of ecological thinking because as a musical form it is a system of improvisation conducted within constraints. Musicologist Martin Clayton contrasts it in this respect to the Western symphony form, which is "the gradual revelation (through linear time) of a pre-existent structure." Each raga is instead a "generative principle" establishing the "parameters" within which an improvisation is undertaken by each musician "according to his temperament."[42]

In his book *Time in Indian Music*, Clayton also asserts that "Indian *rāg* music embodies in some sense a world-view, representing in audible form metaphysical ideas," among them that of "cyclical" time. In the history of global music, the European approach to melody and chord progression is extraordinarily linear. Western music typically creates a narrative curve, which metaphorically parallels the life adventures of a protagonist in a story, ending in closure. By contrast, in raga, the reiterative structure of the melody plays against the relative stasis provided by the drone of the tambura. Instead of providing a narrative progression, "processes . . . unfold continuously within the framework of cyclical time," embodying "both recurrence and change." The opening improvisation (*ālāp*) introduces the raga's structural parameters through a slow process of differentiation among individual tones, from which the "melodic system" emerges. Cycles of inexact repetition ("the same in type, but not in detail") characterize the melodic and rhythmic improvisation that is then built by the performers within the prescribed constraints. For the listener, Clayton writes, "since there is no . . . expectation of . . . final cadence, the ideal condition is . . . being absorbed in an ongoing state of *rag*-ness." After some indefinite duration, the raga ultimately subsides back into the initial undifferentiated drone. Indian musical terminology emphasizes the unfolding reiteration and permutation in the raga as a synecdoche of natural processes, since it is called *barhat*, which denotes a process of organic growth and blooming. In performance, the music develops a structure and becomes a unique entity with affective power, like a living organism.[43]

Other experimental composers of the decade in addition to Young and Riley also absorbed the lessons of classical Indian music. Some studied with Pandit Pran Nath as well, although they did not formally become his disciples. One of these was Charlemagne Palestine, whose musical career mirrors the trajectory of experimental music's historical development. As a teenager in New York City in 1960, Palestine played bongo drums to accompany oral poetry readings by Beat poets Allen Ginsberg and Gregory Corso. Throughout the sixties, he pursued an exploration of what he called "sacred drone" music, first using the carillon bells at St. Thomas Episcopal Church and, toward the end of the decade, electronic oscillators. His electronic works from this period include *Holy* (1967), *Negative Sound Study* (1969), and *Timbral for Pran Nath* (1970). These pieces reproduce as "sacred drone" many of the sounds that were most often vilified as noise pollution, such as the screech of subway trains and the roar of jet engines. In 1970 Palestine left New York City to study at the California Institute of the Arts, where he built his own synthesizer that he called the Spectral Continuum Drone Machine. At CalArts, Palestine also began producing carillon-like overtones on a Bösendorfer Imperial grand piano, reminiscent of the "cloud" sections of La Monte Young's *Well-Tuned Piano*. He would hit the piano's keys in response to emerging overtones, "creating still more overtones, until the effect turns into a kind of natural acoustic feedback loop." In this "strumming" technique, Palestine's listening and his voicing merged dynamically in a complex, self-regenerating drone.[44]

Evolving Pieces: Music as an Ecological System

The creative work of experimental composers was focused on devising open systems that would let sound function as a medium of feedback in nature, thus bringing nature to "hear itself think," while including the human subject as a subsystem in that nature. Their pieces created feedback loops between human performers and their physical environments. Alvin Lucier used "echoes" and "room resonances" in his work to "try to put people into harmonious relationships with" natural phenomena. He contrasted his ecological approach to that of the acoustic ecologists, saying:

> I like to study environmental systems . . . and that's the wrong word . . . systems that occur in the natural world, right? And I don't tape record them. I don't get, go and gather the particular sounds. In other words it doesn't interest me to take a taping machine and tape record the sounds of birds or bats. But what I do enjoy is to study the means by which these animals or the natural world use particular sounds with which to survive. . . . In imitating the way that the natural world works, you find out about it, and you also connect to it in a beautiful way. You don't exploit it. I would feel that tape recording dolphins or bats or something was somehow exploiting their art. I would rather do what they do, on the level that we're able to, you see? That's the difference.

Lucier's composition *I Am Sitting in a Room*, from 1969, uses repeated iterations of a spoken statement to evoke a space's resonant frequencies, allowing the room to "speak." The performer's statement is played on one tape recorder, while another one records; the new recording is then played, and another recording is made of that playback. Slowly the room's acoustics transform the spoken words (including the rhythm of Lucier's stutter) into droning harmonic resonances. *I Am Sitting in a*

Room was first performed for an audience in 1970 and was released as a 45-minute-long recording in 1981. The text reads:

> I am sitting in a room different from the one you are in now. I am recording the sound of my speaking voice and I am going to play it back into the room again and again until the resonant frequencies of the room reinforce themselves so that any semblance of my speech, with perhaps the exception of rhythm, is destroyed. What you will hear, then, are the natural resonant frequencies of the room articulated by speech. I regard this activity not so much as a demonstration of a physical fact, but more as a way to smooth out any irregularities my speech might have.

Each iteration lasts approximately 82 seconds. By about twenty minutes into the performance, all of Lucier's words have passed beyond the threshold of audibility except for his stuttering sibilant "s." For the next fifteen minutes, the resonance pattern, which is all that is now audible, becomes progressively simpler as it is "smoothed out" by repeated iterations. It begins to sound like wind blowing through pipes and wires, with longer sustained notes becoming increasingly prominent. By the end of the piece, there are only three notes recurring insistently (the tonic, the fifth, and the octave) accompanied by a higher-pitched drone.[45]

In the process of these repeated iterations, what was originally the interference or "noise" caused by the resonance of the room has been transformed into the music. As this happens, Lucier's thetic self (the speaking "I" manifested in what he calls the "irregularities of my speech") is subsumed into an environmental resonance—"the natural resonant frequencies of the room"—which is always latently present but has now been made audible, activated or *"articulated by* [Lucier's] speech." The result is that the physical environment "speaks" through, and with the help of, his human voice, which is itself subsumed in the interaction—not gone but "smoothed out" into the ambient drone.

Max Neuhaus's 1977 installation *Times Square* electronically mixes the resonant tones produced in a subway ventilation shaft by passing traffic, to emit a single droning note. Like an electronic version of Pauline Oliveros's simultaneously listening and voicing human subject, Neuhaus's system intervenes in the soundscape by producing a sound after assimilating and integrating all the sounds it has picked up on. The result is a drone that changes with its aural surroundings. The system's voiced tone, selected from among the tones that surround it, turns their "noise" into music by giving the listener's attention a reference point. A review of the work that appeared in the *New York Times* in November 1977 emphasized this process:

> Just south of the statue of Father Duffy at 46th Street between Broadway and Seventh Avenue is a large subway grating on a pedestrian island. If you walk over it, you will hear a steady sound, not unlike that of a conch shell held up to the ear. If you stand still for a long enough time, you will notice that the sound changes gradually and subtly. This is the latest "composition" by Max Neuhaus, who delights in creating environmental aural works. It is the first in a series of Underground Music(s), and it was designed to fit with the sounds around it and yet be different from them. It does not pollute the air nor force itself on one, the way Muzak does in restaurants and other public places, but rather steals into one's consciousness, and by its uniqueness it makes one listen to the sounds. . . .[46]

Neuhaus first discovered how to build such electronic systems in 1963, while realizing a performance of John Cage's piece *Fontana Mix*. In that performance, he used Cage's score to determine when to vary the power settings for four tracks of a sound mixer that was part of an apparatus Neuhaus invented. The mixer governed the relative amplitudes of electronic inputs from four microphones. Each microphone rested on a drum, which resonated in response to the sounds coming from the loudspeakers receiving signals from the mixer, thus creating a system of feedback loops. To Neuhaus's initial surprise, the sound that this generated was not screeching, as he had expected, but complex standing waves: the feedback loops resolved the indeterminacies of Cage's chance composition into sustained harmonic frequencies that varied as the system "adapted" itself to the changes imposed by the score. As Neuhaus described it:

> I had discovered a means of generating sound which I found fascinating—the creation of an acoustic feedback loop with a percussion instrument inserted inside it. Instead of the usual single screeching tones of acoustic feedback, this created a *complex multi-timbered system* of oscillation.

Neuhaus renamed his piece *Feed*, and he compared its intelligent system to a living organism in its capacity for spontaneously restructuring itself in response to multiple environmental factors: the instruments used, the spatial configuration of the various components, the acoustics and ambient sounds of the room, the quality of the first sounds detected by the microphones, and of course the power levels of the mixer's four tracks. As these changed, "the feedback channels suddenly break into different modes of oscillation; sound seems to swing through the room. . . . It seems something alive."[47]

In 1974 Nicolas Collins, who was Alvin Lucier's student at Wesleyan College, devised a sound-generating system that brought together the ideas of Lucier's *I Am Sitting in a Room* and Neuhaus's *Feed*. Collins's *Pea Soup* eliminated the need in Neuhaus's piece both for Cage's score and for the controlling hand of a performer at the mixing board responding to the score's directed random interventions. He achieved this by using an additional nested feedback loop to shift the mixer amplitudes, which he connected to a sensor responding electronically to any movement in the performance space. As Collins later elaborated:

> I had stumbled upon a remarkably simple electronic network that created a site-specific "architectural raga" out of a room's resonant frequencies. . . . Perhaps the most elegant aspect was the responsiveness of the sound itself: one "played" this system not by twiddling knobs or pushing buttons, but by moving or making sounds within field of the feedback.

The result, he wrote in 1974, was a "responsive field" that in his opinion constituted an interactive sonic intelligence. He instructed performers to "Treat Pea Soup as an alien intelligent being who is attempting to gather information about her environment and its residents."[48]

Another strategy for building intelligent systems of sounds was to use the dynamics of interaction among people, instead of electronic circuitry, to form the requisite feedback loops. Michael Nyman in his 1974 book *Experimental Music* categorized this technique as "People Processes: These are processes which allow the performers to move through given or suggested material, each at his own speed."[49] Contextual cues gleaned from listening to the other performers instruct each performer as to how to voice next. Oliveros used this principle in her 1971 sound meditation "Teach Yourself to Fly." But the most influential version of it was the "modular composition" form that Terry Riley invented in 1964 with his landmark work, *In C*.

In C is composed of fifty-three short phrases or "modules," all in the key of C, played by any number of musicians (although the ideal number is around twenty). The performers play "together" in the sense that they are playing simultaneously and in the same space, so that they are aware of one another's playing; but they do not play in unison. They must play the modules in the order in which they appear in the score, but every performer repeats each module as many times as he or she chooses (including zero times, which is tantamount to skipping it). As a result, of the fifty-three modules, three or four are typically in play at any given moment. The resulting performance creates washes of sound with a slowly shifting tonal emphasis. Most good performances last about an hour. Reviewing the piece's premiere, Alfred Frankenstein wrote that "climaxes of great sonority appear and are dissolved in the endlessness. At times you feel you have never done anything all your life long but listen to this music and as if that is all there is or ever will be, but it is altogether absorbing, exciting, and moving too."[50]

The success of such performances depends on the existence of a degree of diversity among the performers, which is organized by the composition's feedback dynamics to create each rendition of the piece. This creates a soundscape ecology akin to murmuration (the patterns emergent in moving flocks of birds and schools of fish). Oliveros wrote in 1972, "Terry Riley's *In C* is like a flock of migrating birds in flight"; and Riley himself likened it to "formations of patterns that were kind of flying together. That's how it came to me. It was like this kind of cosmic vision of patterns that were gradually transforming and changing." Murmuration is produced because the sound "environment" of each performer is constituted by the voices of all the others, making each performer into what Arthur Koestler called a "holon." They are all listening to each other as they simultaneously decide how to voice. The performance challenge posed by the piece, Riley noted, is not that of technical proficiency, but of "how to stay together." Feedback loops of listening and voicing guide the choices of each musician, allowing the presence of indeterminacy without resorting to blind chance. As Riley explained, "I didn't want to have a conductor or someone who was telling the musicians what to do. I wanted them *making their decisions based on their listening*."[51]

After *In C*, the most influential work using modular composition in what Nyman called a "people process" was "Paragraph 7" from *The Great Learning* by English composer Cornelius Cardew, first recorded in 1971. As distinct from *In C*, the murmuration in this piece occurs primarily in reference to pitch, not time. The piece was written for Cardew's Scratch Orchestra, which combined trained and untrained singers—among them, both Michael Nyman and Brian Eno. It includes twenty-four short lines of text. In performance, each line is repeated for a specified number of times, on a varying single note chosen by each singer. A singer chooses the note on which he or she will sing each line according to constraints specified by the score: it must be a note that he or she can hear being sung by a colleague, with the exception that no one can sing the same note on two consecutive lines; and with the further instruction that "if there is no note or only the note you have just been singing, or only notes that you are unable to sing, choose your note for the next line freely." Furthermore, in the interval between singing one line and the next, any singer can move to a new location onstage, putting her voice and ears in proximity to a different set of fellow performers. Nyman described in *Experimental Music* how, as singers move around the performance space getting and giving tones "in a perpetual slow-motion relay procession," the piece's initial cacophony resolves into a chord that is at once both coherent and complex.[52]

Like the evolving self-organizing systems described by general systems theory, these experimental compositions make diversity and indeterminacy into the basis of creativity. The open musical systems that they produce are invigorated by those two factors and governed by feedback, which organizes them into a coherent whole. The pieces enact system dynamics in which the physical environments, social relationships, and the physiologies and levels of training of their various performers might all play a role. Brian Eno explored how this worked in a 1976 essay that he titled "Generating and Organizing Variety in the Arts." "*Somehow a set of controls which are not stipulated in the score arise in performance*" autopoietically, he emphasized.[53]

In his analysis of the dynamics of Cardew's "Paragraph 7," Eno observed that the emergent soundscape resolved into a complex chord that "tends to revolve more or less harmonically around a drone note" that is a resonant frequency of the performance space. The "unreliability" of the untrained singers in the Scratch Orchestra contributes to the piece's musical dynamics, both by introducing mutations through the indeterminacy of the "missed" pitch, and because of their herd-instinct unwillingness to choose discordant notes when faced with the opportunity to sing any note they choose. These factors are key to the success of the murmuration, the first by introducing variety and the second by curbing it. Eno likened the piece in this way to ecological systems in which diversity and symbiosis are both assets, comprising "evolution's way of preparing for a range of possible futures." He would later situate his own sound experiments similarly "on the edge between improvisation and collaboration," where collaboration operates as a constraint on improvisation's potential for creating diversity.[54]

Following the breakthrough of *In C*, Riley moved forward creatively by emulating its murmuration dynamics in a solo performance technique called "live looping." Looping involved splicing together the two ends of a tape recording so that it reiterated indefinitely and, when passed across both the playback and recording heads of a tape player, "allow[ed] a sound to be repeated in an ever-accumulating counterpoint against itself." Riley looped recordings of himself playing simple modules, and overlaid them to produce a moiré of sound, supplementing that with improvised real-time variations. He compared the resulting music to the shapes created by river water running over rocks, in that it had "a changing and movable, rather than merely a strictly repeating, pattern." His first LP release, *Reed Streams* of 1966, consisted of two live-looping pieces, "Untitled Organ" (1966) on the keyboard and "Dorian Reeds" (1964) on the saxophone. Live looping of increasing complexity formed the basis of all his subsequent work. In the seventies, he integrated the structural principles and rhythmic complexity of raga music into his live looping, to create what Keith Potter characterized as "extended and constantly evolving polymetric structures." He primarily used a keyboard in just intonation in these live-looping sessions, recordings of which were released as *Persian Surgery Dervishes, Descending Moonshine Dervishes,* and *Shri Camel.* Like the raga form, these pieces build slowly in complexity and ultimately subside into a drone. Their aural figure/ground relations persistently shift, as new permutations at first call attention to themselves and to the familiar patterns with which they contrast, then gradually fade from attention and merge into the piece's complex soundscape.

Terry Riley's live performances and recordings developed a niche audience, but it was Philip Glass and Brian Eno who brought the form of experimental music to the attention of mainstream listeners beginning in the mid-1970s. Glass, who composed the soundtrack for Godfrey Reggio's *Koyaanisqatsi,* was a prolific composer of pieces in the style of Riley's modular and live-looping performances.[55] His *Einstein at the Beach* was the surprise hit of the 1976 New York opera season. As a result, he received numerous commissions, even composing a piece for *Sesame Street.*

By the time he adopted the experimental music form, Brian Eno was already something of a celebrity as a former member of the successful glam rock group Roxy Music. In developing his "ambient music," Eno built on influences that he had encountered in art school in the sixties, particularly works of La Monte Young and Steve Reich. He also read Gregory Bateson's *Steps to an Ecology of Mind* and other texts on systems theory. As a result, he was intrigued by the possibility of making music that could be a "self-regulating and self-generating system." The 1975 album *Discreet Music* was his first sustained engagement with the experimental form. *Discreet Music* was composed of multiple "phased loops": looping modules overlaid on one another while being played at different speeds. It was intended, he wrote in the liner notes, to be played at the threshold of audibility, so that it would mix imperceptibly with the listener's ambient sound environment. He first applied the term "ambient music" to his 1978 *Ambient 1: Music for Airports*.[56]

Brian Eno's Ambient Music for Airports

In the liner notes to *Discreet Music*, Eno described the (likely apocryphal) epiphany that had led him to "invent" ambient music. He was bedridden and unable to turn up the volume on an album of baroque harp music that was playing on his stereo. The almost inaudible music became integrated with the other sounds in the room, producing a new kind of music that did not dominate his attention but existed "as part of the ambience of the environment just as the colour of the light and the sound of the rain were parts of that ambience. It is for this reason that I suggest listening to the piece at comparatively low levels, even to the extent that it frequently falls below the threshold of audibility." The result, he hoped, would be "a music that included, rather than excluded, [and] that didn't have a beginning or an end," but that was continuous with its surroundings as if it were "just sounds that happen to be drifting through the atmosphere at that time." Instead of coming to a definite conclusion, the music fades out so as to leave the impression that "it is *continuing out of earshot* . . . a section from a hypothetical continuum."[57]

In *Ambient 1: Music for Airports*, Eno purposefully engaged the discourse of noise pollution, invoking the identification of airports with the epitome of environmentally harmful noise. He positioned his ambient music as an alternative to the Muzak that was typically used in airport settings to provide what Murray Schafer had termed "audioanalgesia," the masking of the actual environment with distracting, irrelevant sounds. Eno wrote, "Whereas the extant canned music companies proceed from the basis of regularizing environments by blanketing their acoustic and atmospheric idiosyncrasies, Ambient Music is intended to enhance these." As opposed to the dominant strategy of shutting out a potentially annoying sonic environment, ambient music intended to weave the ambient noise together into a music of the immediate environment. "Most importantly for me, it has to have something to do with where you are and what you're there for," Eno stated.[58]

Eno reiterated the claims made by earlier pioneers of experimental music that the form was intended to serve a social purpose and was not mere entertainment. He wrote, "Composition of this kind tends to create a perceptual shift in a listener." It "suggests, by its own internal mechanism, a new way of dealing with the environment, of becoming reoriented." This reorientation aimed to facilitate the listener's reconnection with the natural environment and its living systems. As Eno explained in an interview, "I look at the variety of the world and of the organisms and so on within it and instead of saying, each one of these is an entirely separate phenomenon, I say, each one of these is the product of quite a small number of forces and constraints, reconfiguring in different ways." In particular he elaborated on the perceptual shift that he hoped to produce with his music for airports: "I want to make a kind of music that prepares you for dying—that doesn't get all bright and cheerful and pretend you're not a little apprehensive, but which makes you say to yourself, 'Actually, it's not that big a deal if I die.'"[59]

SIX

Dancing with Animals

Real riding is a lot like ballroom dancing or maybe figure skating in pairs. It's a relationship. TEMPLE GRANDIN, *ANIMALS IN TRANSLATION*

The idea that intelligence took forms other than conscious thought was fundamental to ecological thinking. According to general systems theory, thinking was not the activity of a transcendent intellect, but immanent in the responsive processes of open systems to the larger systems in which they were nested. The physical body's kinesthetic and sensory pathways were thus key participants in human mental processes. Besides forming the basis for consciousness, they gave rise to an intuitive grasp of the environment that registered as "feelings." Such feelings, arising from the responses of the somatic and autonomic nervous systems, were believed to represent a more inclusive and holistic intelligence than consciousness could provide. The experience of empathy constituted communication on this level of intelligence.[1]

Because ecological thinking privileged feelings and empathy over consciousness and language as the means of cognition and communication, it left little basis for claiming that humans held a monopoly on intelligence. While it was still radical to imagine the intelligence of plants, numerous popular works from the seventies advocated the recognition of animals as intelligent because they were feeling and communicative beings. In the introduction to *Mind in the Waters: A Book to Celebrate the Consciousness of Whales and Dolphins*, published in 1974 with the help of the Sierra Club, environmental activist Joan McIntyre declared, "We have, for too long now, accepted a view of non-human life which denies other creatures feelings, imagination, consciousness, and awareness." In the book *Animal Liberation*, published in 1975, philosopher Peter Singer also argued that from an ethical standpoint "All Animals Are Equal" on the basis of their capacity to respond emotionally to their environments. "Although humans have a more developed cerebral cortex than other animals," Singer wrote, "impulses, emotions, and feelings are located in the diencephalon, which is well developed in many other species of animals, especially mammals and birds." In Singer's view, the widespread notion that the definitive measure of intelligence was a capacity (or incapacity) to learn human language was absurd. It merely served the ideological purpose of justifying human exploitation of other animal species. Even people, Singer noted, did not typically use such verbal language to express their feelings: "We tend to fall back on nonlinguistic modes of communication such as a cheering pat on the back, an exuberant embrace, a clasp of the hands, and so on." Many animals, he wrote, communicated their feelings of pain, fear, anger, love, joy, surprise, and sexual arousal by using a similar paralinguistic repertory that, if properly appreciated, should evoke even interspecies empathy.[2]

In conjunction with the growing impact of ecological thinking and its emphasis on empathy, the seventies witnessed a new focus on the affective quality of human-animal interactions. Jane Goodall's *In the Shadow of Man* described how she established rapport with a group of chimpanzees in Tanzania. The Monks of New Skete developed an empathy-based dog training technique that they published under the title *How to Be Your Dog's Best Friend.* The San Francisco–based art collective Ant Farm received funding from the Rockefeller Foundation to design a floating Dolphin Embassy that would include "interspecies living rooms" where humans and cetaceans could relax and play together. Temple Grandin began redesigning cattle yards as a result of imagining them from the animals' point of view.[3]

Acknowledging the emotional lives of animals demanded moving beyond behaviorist approaches to animal behavior, which remained rooted in the dualism of mind and matter that characterized Enlightenment science. This led to a particular excitement about exploring new forms of human relationship with horses and small toothed whales (dolphins and killer whales), as these were two groups of animals that were known to resist behavioral conditioning. Due to its reliance on empathy and physicality, the new ideal for interacting with animals was often described in ecological texts as a kind of dance. Gary Snyder said that to "look into an animal's eyes and see an intelligence there, a sensibility," was an aspect of the impending American rebirth into nature that he called "the ecstasy of the dance."[4] A range of innovative cultural practices including flotation therapy, contact improvisation, and horse whispering fostered empathic connections between humans and animals by taking the notion of dance more literally, as interactive bodily movement.

The goal of achieving better emotional rapport with animals was not entirely directed toward establishing their more humane treatment; it was also predicated on the idea that humans could benefit from what animals could teach them. Goodall wrote that she learned effective child-rearing practices from chimps. Grandin built a version of the squeeze chute that calmed cattle and used it to calm herself.[5] Establishing a better rapport with animals ultimately included establishing better relations with one's own animal self. In the dance form known as "contact improvisation," developed in 1972 by Steve Paxton, the primary dictum was to "take care of your animal." According to Daniel Lepkoff, who studied with Paxton in the early seventies, "'One's animal' is a physical intelligence . . . that form[s] our ability to survive and to meet and play energetically with our environment. A main aspect of the early Contact Improvisation work sessions was to coax, encourage, and engage this animal intelligence." This was accomplished, Lepkoff noted, mainly by "getting out of its way, letting go of one level of control and learning to trust in another": a decentralized kind of control created by physical interactions with another, in a feedback relationship that put "body experience first and mindful cognition second."[6]

Beyond Behaviorism

The environmental movement of the seventies voiced repeated criticisms of the treatment of animals in mainstream American culture. Farley Mowat, author of the memoir *Never Cry Wolf,* in 1972 published another work, titled *A Whale for the Killing,* which became a best-selling paperback in 1975. *A Whale for the Killing* sparked public outrage that galvanized Greenpeace's nascent "Save the Whales" campaign. In it Mowat described how the inhabitants of a fishing village in Newfoundland slowly killed (over several days, using hunting rifles, handguns, and speedboats) an eighty-ton pregnant fin whale trapped by the tides in their shallow harbor.[7]

In his melancholy conclusion, Mowat suggested that the townspeople had reacted defensively against the uncanniness of the whale's subjectivity. Its sublime, alien presence in the familiar waters of "their" port evoked a rabid antipathy because of what it threatened to reveal to them about themselves. As Mowat wrote, "An awesome mystery had intruded into the closely circumscribed order of our lives; one that we terrestrial bipeds could not fathom, and one, therefore, that we would react against with instinctive fear, violence, and hatred."[8]

Consistent with Masahiro Mori's theory of the "uncanny valley," the psychological defenses against another being's uncanniness typically take two forms: distancing it, even to the point of destroying it; or exaggerating its likeness in order to claim it as a dependent mirror image of ourselves. Historically, humans have exhibited both strategies in their attitudes toward animals. Pets are often treated like permanent children; while whales in particular have been cast as monsters and consequently killed with callous abandon. Until 1957, Mowat related in *A Whale for the Killing,* U.S. Navy aircraft routinely used whales as opportunities for target practice in anti-submarine warfare.[9] The ape is another uncanny monster of long standing. As Paul Shepard observed in *Thinking Animals,* "It needles us because the ape is uncomfortably close as an animal and disgustingly far away as a human."[10] The strangely bifurcated treatment of apes by American scientists in the sixties and seventies exemplifies both defensive responses.

Some scientists of the era tried to set aside the difference between apes and humans and raised baby chimpanzees at home with their families. Wearing diapers and living in a house, learning American Sign Language in the same way that a human child would learn a primary language, these apes were constructed as surrogate humans. The most famous among them were Lucy, Washoe, and Nim. Their upbringing was intended to test the extent of their capacity for humanization.[11]

However, seen in this way as an "almost" human dependent—a human subject in a Lacanian fun-house mirror, so to speak—the baby chimp in diapers was always already an image of inadequacy or incompleteness: a falling short of the truly human, rather than a truly animal intelligence. When Jane Goodall's pioneering study of chimps in the wild, *In the Shadow of Man,* appeared in 1971, David Hamburg of the Stanford University School of Medicine felt compelled to pen a foreword in which he characterized chimps as "capable of close and enduring attachments, yet nothing that looks quite like human love, capable of rich communication through gestures, postures, facial expressions, and sounds, yet nothing that looks quite like language."[12]

The predisposition to see these chimps as always-already-not-quite human shaped the course of the language studies in which they were enlisted; these ultimately conformed to the master narrative of inadequacy and failure. Relying on a very restrictive definition of "language," the investigators consistently hung their learning objectives just out of the chimps' reach. Nim learned a vocabulary of hundreds of signs and combined them spontaneously to converse with his human keepers; but because his utterances lacked grammatical syntax, Herb Terrace, the principal investigator, concluded that Nim and chimps in general were incapable of acquiring language. By "disproving" the claim that chimps were capable of language, Terrace succeeded professionally exactly where Nim failed.[13]

As the subjects of scientific experiments in the seventies, while some chimps wore diapers and struggled to learn sign language, others were hooked up to machines, exposed to electrical shocks, and dismembered. The Cartesian dualism intrinsic to Enlightenment science was manifested most baldly in the discursive parameters of behavioral ethology. Animals, according to Cartesians, were without consciousness; they only responded mechanically to physical stimuli. Behaviorism likewise rejected on principle the imputation of emotions to animal subjects, on the basis that parsimony (known colloquially as "Occam's razor") precluded attributing such states of awareness to explain behaviors for which simple stimulus-response explanations would suffice.

According to ethicist Peter Singer, behavioral ethologists who were engaged in primate research were in denial of a contradiction at the core of their work. Its potential usefulness was based on the premise that apes were much like humans. But the availability of apes as research subjects was based on the opposite premise that they were not. This contradiction was foregrounded in Frederick Wiseman's 1974 documentary, *Primate,* which took viewers on a virtual tour of the Yerkes Regional Primate Research Center at Emory University. The film's opening scene confronts viewers with a disturbing disparity. Staid portraits of esteemed scientists in a hall of honor at the Center suggest the dignified pursuit of beneficent knowledge. But the accompanying sounds of slamming metal gates and echoing animal cries evoke associations of abusive prison guards and tortured inmates.

Wiseman's film enlists viewers to see the workings of the Primate

Research Center as exemplary of science's failure of empathy. We are asked to watch as scientists implant electrodes in the brains of unwilling apes and monkeys and subject them to a variety of electrical stimuli. One scientist in an interview explains how ape behavior is studied using a checklist, with columns for twenty categories of behaviors, and rows for minute-length time intervals; his engagement with apes takes the form of quantifying their "frequency of responses per unit time." The techniques of emotional distancing reach an absurd extreme as the scientist speaks into a tape recorder while watching a mother ape with her newborn infant. The viewer sees her hugging and comforting her infant in a very recognizable maternal gesture, while the ethologist restricts himself to a reductively objective description of the event couched in behaviorist jargon: "The infant vocalizes . . . it stops vocalizing . . . the mother rolled onto her back briefly; she patted the infant clinging to her *ventrum*."

Ecological thinkers of the 1970s sought to replace behaviorism with other ways of learning from animals that did not reduce them to objects. "Cognitive" ethology recognized the existence of what might be called the "animal mind." Its logic reversed the parsimony argument of behaviorism, asserting that "since cognitive theory holds for humans, it is unparsimonious not to apply it to animals." Animals and humans shared a common evolutionary history; it was unjustifiably complicated to ascribe different theories of mind to each. This position was echoed by Singer, who wrote in *Animal Liberation*:

> The nervous systems of other animals were not artificially constructed to mimic the pain behavior of humans, as a robot might be artificially constructed. The nervous systems of animals evolved as our own did, and in fact the evolutionary history of humans and other animals, especially mammals, did not diverge until the central features of our nervous systems were already in existence. . . . It is simpler to assume that the similar behavior of animals with similar nervous systems is to be explained in the same way than to try to invent some other explanation for the behavior of nonhuman animals as well as an explanation for the divergence between humans and nonhumans in this respect.[14]

Some cognitive ethologists even began to insist that encountering animals wild and "in the field," as opposed to in the laboratory, was the only way to learn anything real about them. There was a growing awareness that some among them—often "amateurs" with minimal professional indoctrination when they started their careers—had already taken this radical course: Jane Goodall with chimps; Farley

Mowat with wolves; and Ken Norris, Karen Pryor, and Jacques Cousteau with dolphins. Cousteau wrote in 1975:

> It is our own firm opinion that present studies of the intelligence and the souls of dolphins are violated by the conditions in which they are carried out. . . . We have succeeded in creating a personality common to captive dolphins. And it is that personality that is being studied, without taking into sufficient account that we are dealing with animals that have been spoiled and perverted by man.

The normally equanimous Goodall also wrote toward the end of her book:

> Most people are only familiar with the zoo or the laboratory chimpanzee. This means that even those who work closely with chimpanzees, such as zookeepers or research scientists, can have no concept or appreciation of what a chimpanzee *really* is.[15]

Marine biologist John Lilly was one defector from the ranks of professional scientists to adopt this way of thinking. Lilly had criticized the laboratory methods used on animals as early as 1958, when he published results from his *in vivo* electrical stimulation of monkeys' brains (the kind of work later documented in *Primates*) in which he mentioned parenthetically that he intuitively felt that monkeys had feelings, regardless of his professional training to the contrary: "When an intact monkey grimaces, shrieks, and obviously tries to escape, one *knows* it is fearful or in pain or both." In 1963 Lilly ingested LSD for the first time. Under its influence, he felt himself to be a unit of information in a vast system, like "a single thought in a huge mind, or a small program in a cosmic computer." He reported, "I came back to my body from that experience with full respect for the possible varieties of life forms that can exist in the universe . . . the varieties of intelligences that exist." At the time that he had this psychedelic insight, he was running dolphin research laboratories in Miami and St. Thomas funded by the National Science Foundation and the U.S. military. But in 1966 he discontinued his research until it could be reestablished using dolphins that were not confined in tanks and compelled to participate.[16]

In the book that Lilly published the following year, titled *The Mind of the Dolphin,* he maintained that animal behavior and communication could not be studied as if they occurred in a contextual vacuum. Because animals were sentient subjects, the context in which their behavior was studied was all-important; and the laboratory in this respect constituted the diametric opposite of the research context that was necessary. True knowledge of animals had to be based on a two-way communication between the human researcher and the animal subject. The extent and content of this communication would necessarily depend on the affective quality of their relationship. In short, to learn about animals, one needed to cultivate empathy and learn *from* them. The result would be an affectively charged social interaction between two animals: one human, one nonhuman; and the human researcher would inevitably also be the subject of the animal's investigations. So far, Lilly wrote, in their efforts to learn about animals, humans had not been not putting their best foot forward.[17]

The Willfulness of Dolphins and Horses

The new enterprise of learning through affective relationships with animals focused largely on horses and small toothed whales.[18] While the intelligence of chimps was widely acknowledged, ecological thinking was seldom present in the discourse surrounding ape intelligence, since they were regarded as not-quite-human. The intelligence of dolphins and horses, by contrast, was typically framed as distinctively nonhuman.

The emphasis on learning from dolphins and horses did not derive from their intelligence alone. It was due also to the fact that humans had formed working relationships with both species; and, significantly, the dynamics of those relationships could not be reliably mastered by using only behaviorist techniques. Horses and dolphins responded to "operant conditioning," to be sure; but they also often exceeded it or willfully defied it in ways that forced their humans to acknowledge the animal's "point of view." They seemed to have their own ideas and ways. And because of their size and strength, not acknowledging those could be dangerous. As horse trainer Ray Hunt laconically observed, "They can buck us off, and run us over too." Dolphin trainers and horse trainers who spent large amounts of time working with their animals "in the field" staked out an intellectual middle ground between New Age enthusiasts like John Lilly and laboratory-based ethologists who denied the possibility of animal consciousness.[19]

Dependable behavior from a dolphin or a horse, their trainers had learned from experience, could only be garnered through a two-way communication involving negotiated boundaries and mutual trust. Humans working with dolphins and horses found themselves not *in* control but part of a dyadic system with dual controls. They were required to work within the constraints dictated by the dynamics of that system. Hunt called it "patterns of relating" with "mutual respect and discipline." Pryor called it a game with negotiated rules that both species had agreed to. The horses and dolphins made some of those rules. If the humans strayed outside of the parameters, the animals responded with a range of negative affects, including boredom, laziness, frustration, and disgust. Hunt wrote that to be successful, the horse trainer must "get [the horse] in a learning frame of mind":

> You try to keep things where they are not boring for the horse. You do a variety and variation of different things when you are trying to recognize how to understand the horse's movements. You do not get him in the wrong attitude. You do not want him to feel you're boring him, and that what you're asking is monotonous.[20]

For breaking their rules, dolphins and horses would discipline their trainers. In his 1978 book, *Think Harmony with Horses: An In-Depth Study of Horse/Man Relationship,* Hunt argued that if a rider had trouble with a horse,

> I'll prove to you that the horse is right and we are wrong. . . . [Y]ou have to get discipline within yourself so that you can have it with your horse. If you don't, this is what will cause your horse to get cranky and take over. . . . It's because he knows you don't mean what you're talking about . . . because we are so superior, or neglectful, or lazy. Because we haven't prepared ourselves to recognize the horse's feelings.

Dolphins, for their part, disciplined their trainers primarily by sulking in their tanks and refusing to engage. But Karen Pryor related in a 1973 publication that she had seen a diver, who had accidentally kicked a porpoise when leaving the pool, be met with a blow of equal intensity at the first opportunity the next morning. She also noted that dolphins would splash a human who they thought "should be helping [them] and isn't."[21]

Beyond such specific disciplining behaviors, horses and dolphins discriminated generally among people and decided whom they would and wouldn't work with. In his 1974 book *The Porpoise Watcher,* dolphin trainer and zoologist Ken Norris told of how

> a male trainer who briefly took over part of the training duties never achieved the slightest rapport with Pono. He was harsh, abrupt, and insensitive to her feelings,

using conditioned response training techniques in an austere manner that "turned off" Pono completely. She quickly became refractory and began to nip. Before long he could not come to tankside without Pono's rising up in an attempt to grasp him with her long, efficient rows of teeth. At length I had to relieve him of his duties [and] . . . Pono, within a day, reverted to her usual docile self again.

Norris interpreted Pono's refractory behavior not as disobedience, but as the trainer's failure to negotiate a trust-based relationship. "Pono picked and chose her human friends with care, accepting those who for one reason or another had won her trust, and rejecting those who had not," he wrote. This was a quality that he ultimately admired: "This central core of dignity, of the measure of her own rights and worth, lay deep in Pono's spirit." Hunt similarly explained to his readers that a horse needed to have "confidence" in a rider as the leader. A horse, Hunt warned, assessed a rider's behavior; if one lost his respect, he wouldn't even try to do what was asked of him. "He will question you," Hunt predicted; if in response the horse felt fear or doubt, the horse would decide it was safest not to heed you.[22]

Conversely, the successful establishment of a dyadic relationship achieved levels of communication and accomplishment far exceeding what could be expected under the behaviorist paradigm. Hunt wrote, "These horses are more sensitive than we can ever imagine. . . . You develop this sensitivity. Let them use their keenness to show how sensitive they are—to teach us." Pryor commented in her 1975 book *Lads Before the Wind*, "It is really an eerie thrill when the animal turns the training system around and uses it to communicate with you." She wrote, "I have myself seen porpoises 'misbehave' to get a message across . . . 'saying,' in porpoise body English, 'Hey, Randy, before we go to work, take a look. I've got a piece of wire stuck in my back teeth—take it out, will you?'" Furthermore, she reported, "I have seen an animal test the training situation to ask and answer its own questions." As an example, she described an incident in which she was teaching two false killer whales to jump in tandem. One, named Olelo, kept getting it wrong: she jumped just after the other whale, then in unison but from the wrong side, then almost but not quite in unison. Pryor recounted, "I gave Makapuu her usual two-pound reward and gave Olelo one tiny small smelt. Olelo physically startled and looked me in the eye," and afterward always jumped perfectly. Pryor concluded that Olelo had been testing the amount of leeway that was acceptable and had correctly interpreted the unprecedented reward of a single smelt as a sign for "half-right." She offered this as evidence of an interspecies communication "admitting of spontaneity on both sides" and therefore "capable of both giving and receiving information that is far more subtle than that normally conveyed" in behavioral conditioning environments. "She [Olelo] had been trying the scientific method *on us*, deliberately testing the exact nature of the rules," Pryor averred. "Her testing, in about ten minutes of work, gave her the answers she needed. That's communication."[23]

Intersubjective dolphin-human relationships also exceeded the behaviorist paradigm in eliciting dolphin displays of what Gregory Bateson called "deutero-learning." Bateson spent most of the 1960s studying dolphins—first with John Lilly in St. Thomas and then, after Lilly shut down his operations, with Karen Pryor in Hawaii. Together Bateson and Pryor explored whether dolphins were capable of learning what it meant to learn. Pryor began rewarding two *Steno* dolphins, Malia and Hou, only when they did something new every day that she had never rewarded them for doing before. Both dolphins eventually caught on to this new, more abstract game. Bateson later told Stewart Brand: "Between the fourteenth and fifteenth session the porpoise [Malia] got awfully excited in the holding tank, slapping around. She came on stage for the fifteenth session and did twelve new things one after another, some of which nobody'd ever seen at all in that species. She'd got the idea." As opposed to the slow-paced incremental reinforcements that were considered a necessity in operant conditioning, Pryor's cetaceans could perform a trick after seeing it done only once, having "generalized the concept that sounds are 'cues' and different sounds signal different responses." As an extreme case, once Malia and Hou were mistakenly switched in their tanks; and when it came time to perform, each did the other's tricks, although they had never been formally taught to do them.[24]

A third way in which intersubjective interspecies relationships exceeded the behaviorist model was in their creation of mutual bonds of affection that went beyond the purposive needs of training. Ethologist Konrad Lorenz, who visited Pryor's facility in 1967 in order to spend some time with Bateson, later described a scene that had happened there:

> A female porpoise who (I intentionally say "who" and not "which") had struggled for some time with a learning problem had obviously begun to suffer from her inability to solve it. When, with the help of the trainer, she suddenly succeeded in getting the point, she did something which no porpoise had ever been observed

> to do—she swam up to the trainer and stroked her with her flipper, a gesture of endearing caress which is often observed among friendly porpoises but never had been recorded as being addressed to a human being.

Empathic relations entailed human empathy for the animal; they resulted in a reciprocal animal empathy for the human.[25]

Many writers agreed that these affective relationships helped people to become better human beings. Hunt wrote that developing improved relations with your horse would increase your worthiness as a human. "To understand the horse you'll find that you're going to be working on yourself. . . . [K]now what you have to offer your horse so that your horse can come through for you." Paul Spong, who in 1974 convinced a group of antiwar activists calling themselves Greenpeace to focus instead on saving the whales, said that he was taught how to overcome fear by a killer whale named Skana when he worked for the Vancouver Aquarium in 1968. She taught him by raking her teeth lightly across his feet as he dangled them in the water, repeating the action until he stopped flinching at it. Fired from the aquarium for publicly advocating Skana's release on the grounds that orcas became depressed and bored in small tanks, Spong founded KWOOF (Killer Whale *Orcinus Orca* Foundation) with the help of funding from Stewart Brand. He convinced Greenpeace to devote themselves to the anti-whaling crusade by taking their leader, Bob Hunter, to visit Skana. She gently gripped Hunter's head in her powerful jaws, held it, and let it go. This amounted, Hunter later wrote, to the best encounter-therapy session he had ever had: "I could see exactly who I was, what my limitations were. It was a gift."[26]

Ken Norris concluded his book *The Porpoise Watcher* with the hope that, in the looming face of environmental disaster, affective relationships with animals could convince humans to relinquish their shortsighted purposiveness. Our survival as a species, he wrote, could not be achieved by controlling nature to our own ends through scientific knowledge. It depended instead on cultivating the adaptive tools of the dolphin trainer: an "alertness to change" and "the chance to reverse [course]." This was what working with dolphins had taught him, which he extrapolated to an ethic:

> To tamper gingerly with natural systems, relying upon their own internal integrity over a space of time to reveal the consequences [of our actions] to us. . . . There are no such things as simple causes and effects. Instead . . . the world is composed of such unimaginably complicated interlocking systems so full of feedback loops.[27]

Whale Song

The television show *Flipper*, broadcast on NBC beginning in 1964, followed by a flurry of popular books in the seventies, spread the idea that dolphins and whales were highly intelligent and empathic animals, maybe even telepathic.[28] John Lilly wrote that human efforts to communicate with them were hampered by our own overemphasis on visual communications, as opposed to the acoustic and the haptic.[29] Dolphins were more attuned to the latter two, since light did not penetrate very deeply into the ocean, and they relied on echolocation to navigate.

Among the senses, the ecological discourse as a whole privileged touch and sound as the means of communication most suited to animal intelligence. Paul Spong, captivated by Lilly's idea, brought local musicians in to the Vancouver Aquarium at night to play for Skana; the orca seemed to prefer flute and violin music, sometimes emulating the voices of the instruments. In August 1970, Spong steered a floating soundstage equipped with underwater speakers out into Vancouver Bay to play folk music for the cetaceans there. Other members of Greenpeace played flute and saxophone music to whales.[30]

Whales, though, seemed generally less interested in human song than vice versa. *Songs of the Humpback Whale*, consisting of five tracks amounting to thirty-five minutes of humpback whale vocalizations recorded off the coast of Bermuda, sold over 100,000 copies when it was first released in 1970. Roger Payne, who had collected the recordings, coauthored a scholarly article that was published the following year in the journal *Science*, justifying his use of the term "song" in reference to the whale sounds. Not only was it a song, Payne wrote, but it was an evolving song—different from year to year, but sung the same way each year by all the humpback whales within one another's hearing. The structural similarities of the songs sung by different groups of whales also implied that they were improvised within a shared set of rules. Could this be the whale equivalent of language? Gregory Bateson thought so; he wrote in *Steps to an Ecology of Mind* that the language of a species without hands would not develop a syntax focused on describing the manipulation of objects, as human language had, but rather one that focused on the expression of interrelationships within the school or pod. "We do not even know what a primitive . . . [symbolic] system for the discussion of patterns of relationship might look like," he considered, "but we can guess that it would not look like a 'thing' language. (It might, more probably, resemble music.)"[31]

Over the objections of scientists and linguists who disputed the application of the terms "song" and "language" to whale vocalizations, tracks from *Songs of the Humpback Whale* were entered as testimony during hearings at the Department of Interior in 1970, and again during hearings of the House Committee on Foreign Relations in 1971—almost literally the "speech on the floor of Congress from a whale" that Gary Snyder imagined as a model of environmental consciousness. Those hearings resulted in a congressional resolution calling for the U.S. secretary of state to negotiate a ten-year global moratorium on whaling. At the time, no whale biologist (most of whom worked closely with the whaling industry) supported the resolution, but public opinion was strongly for it. Whale song had captured the public imagination.[32]

Whale song echoed in works of popular and classical music composed in 1970 and '71, when the debate over a whaling moratorium was regularly in the headlines. Folk singer Pete Seeger wrote a protest song called "The Song of the World's Last Whale." Avant-garde composer George Crumb directed a blindfolded flautist to imitate whale song in his trio *Vox Balaenae for Three Masked Players*. Experimental composer Alvin Lucier wrote *Quasimodo the Great Lover*, which emulated the whales' method of "long-range sound-sending."[33] Alan Hovhaness sampled Payne's actual recordings in his orchestral piece *And God Created Great Whales*. Judy Collins used them too, in the song "Farewell to Tarwathie" on her *Whales and Nightingales* album, which sold over 250,000 copies.

As the political struggle to extend and enforce the whaling moratorium went on, works integrating humpback whale song continued to be made through the end of the seventies. David Crosby and Graham Nash reached number six on the *Billboard* charts with their 1975 *Wind on the Water* album, featuring a song titled "To the Last Whale," which used Payne's recordings to supplement lyrics ascribing feelings to the whales. Paul Winter's 1978 album of New Age music, *Common Ground*, included recorded whale song on two numbers: "Ocean Dream" and "Trilogy," which mixed it with the howl of the timber wolf and the cry of the bald eagle. In 1979 jazz bassist Charlie Haden released "Song for the Whales" on the second *Old and New Dreams* album, with saxophonist Dewey Redman imitating whale song on a musette.

A second album of Payne's recordings was released by Capitol Records in 1977, under the title *Deep Voices: The Second Whale Record*. In that same year, NASA sent humpback whale song out into deep space on the Voyager 1 and Voyager 2 missions, as an example of the best sounds that Earth had to offer. In 1979 *National Geographic* distributed ten and a half million copies of recordings from Payne's second album in a 45-rpm magazine insert. The insert accompanied an article by Payne that described his experience of sailing at night off the coast of Bermuda, feeling profoundly lonesome until he lowered his hydrophones and found companionship in the presence of the whales, whose underwater songs "wove together like strands in some vast and tangled web of glorious sound."[34]

As Payne's words suggest, the popular discourse around whale song took it as sublime evidence of an intelligence that might be superior to our own. Paul Spong of Greenpeace called whales "serene super beings in the sea." Inverting the traditional representation of whales as uncanny monsters, he imagined them as "a nation of Buddhas" confronted by the "vicious monsters" that were humans: "whose only response to the natural world was to hack at it, smash it, cut it down, blow its heart away."[35] This perspective was shared by the 1977 film *Orca*, an ecologically minded answer to the 1975 blockbuster *Jaws*. In *Orca*, a scientist who is in Newfoundland to record whale song comes into conflict with a collector with harpoon guns who wants to catch a killer whale to sell to an aquarium. The film uses whale song as an affective tool in conveying its message that the "killer whale" should be an object of empathy as well as of fear. The opening credits use orca vocalizations blended with an oboe melody to evoke a sense of sublimity. This soundtrack accompanies images of two leaping orcas presented as a happy couple in the wild.

The movie offered its viewers a mix of science and New Age myth in portraying orcas as beings of super-human intelligence. It constructed the orca's violence as an instance of nature's feedback to humans: a communication intended to discipline us for exceeding the constraints of a previously established symbiotic relationship. In the first scene, a man is pursued by a great white shark, which, before it can reach him, is killed by an orca who races to the rescue. We are then brought to sit in on a college lecture on the killer whale, in which the students are informed that "treated with kindness, there is no creature that is a greater friend to man. . . . Yet the most amazing thing about these creatures is neither their gentleness nor their violence but their brains," which indicate an intelligence far superior to ours (fig. 10). Whale song is presented as additional evidence of this higher intelligence. We and the other students are told that "analyzed by computers at Caltech,

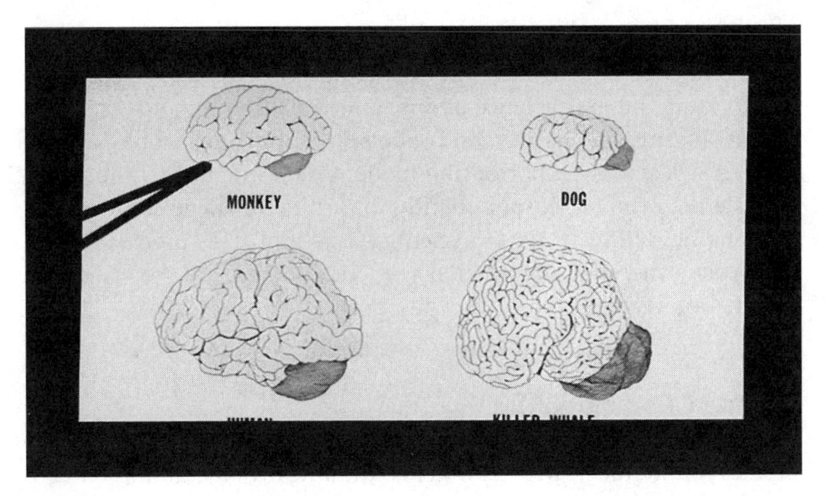

Figure 10. Comparison of animal brain sizes in *Orca*. The human brain is bottom left, the orca's bottom right.

[whale song] was found to contain fifteen million pieces of information; the Bible contains only four million." As compared to whale song, the lecturer says, "What we call language they might call unnecessary, or redundant, or retarded."

The movie is not subtle in driving home its moral that humans are the real monsters of the seas. Although an orca kills a great white shark to save a human's life, this altruism is not reciprocated. The whale collector instead kills the orca's pregnant mate while attempting to catch her. The orca therefore pursues him and his crew relentlessly for vengeance, like a hybrid of the great white shark in *Jaws* and Charles Bronson's character in the 1974 film *Death Wish*. As the violence escalates, several other characters berate the collector until his original attitude of cocky superiority is replaced by guilt and remorse. The scientist tells him that the orca is "a fellow creature, like you . . . with intelligence." An Inuit (played by Native American actor Will Sampson) corroborates, "She speaks you the truth. She knows it from the university. I know it from my ancestors." The local priest answers in the affirmative when the collector asks doubtfully, "Can you commit a sin against an animal?" and counsels him, "Sins are really against oneself."

When the collector is at last converted to this point of view, it is revealed that his previous recalcitrance was a psychological defense against the uncanny similarity of the orca's story to his own. "The same thing happened to me," he confesses to the scientist: a drunk driver accidentally killed his pregnant wife. His defenses dissolved, he now feels great empathy for the whale. But his realization comes too late to avert the malevolence of a wounded nature; the orca has been driven "insane." It kills everyone onboard the ship one by one—a natural feedback system in horror film mode. Late in the movie, the collector listens to the orca's song nightly in a ritual of tragic communion. "What's he telling you?" the scientist asks him. He answers, "'Well, now you're me,' he's saying. 'I'm you,' he's saying. 'You're my drunk driver,' he's saying."

Animal Choreographies

Besides sound, touch was the other communicative medium privileged by the ecological discourse in the expression of animal intelligence. Haptic perception was not conceived to be entirely separate from sound. Sound waves physically impact the body. Infrasonic frequencies, which are a key element of whale song, cause visceral vibrations in humans even though they cannot be detected by the human ear. Ecological writers insisted that many animals' senses were not as discrete as was implied by the received notion of five different senses, each localized in its specific receptor organ. British zoologist Maurice Burton suggested in 1972 that what was popularly referred to as a "sixth sense" that animals seemed to possess connoted an ability to integrate multiple diffuse sources of sensory information into a unified "proprioception" that "registers less obvious stimuli than do the specialized senses, and has usually been held responsible for what is called kinaesthetic sense." Supporting this theory, Ken Norris established through a series of experiments that dolphins "hear" with their jaws and heads, as well as with their ears. Acoustics for them is part of a broader haptic perceptiveness that is not localized in specific organs. Norris marveled in 1974, "How a porpoise might hear with its forehead escapes me to this day. No acoustic nerve penetrates that area, and there seems to be no route for sounds hitting there to get directly to the ear."[36]

Understanding proprioceptive responsiveness as a kind of intelligence problematized the customary distinction between smarts and skills. Dolphin trainer Ric O'Barry recalled arguing this point at the Miami Seaquarium with an anthropology student who questioned the concept of cetacean intelligence. He challenged her to jump up out of the water and take a cigarette from his mouth as the dolphin named Clown had just done in the show. When she admitted that she couldn't, he told her, "You just failed the Dolphin IQ test," since their "skill is a form of intelligence." The cigarette jump involved not just physical strength, O'Barry pointed out, but "jumping exactly. Jumping at an exact place and touching an exact spot at an exact moment." This could only be accomplished as a result of intricate calculations—a truth that was demonstrated when once Clown made a mistake and bit off part of his cheek. "If [the dolphin] didn't pay attention to the details of the jump, if she couldn't concentrate on what she was doing, there would be no cigarette jump. And what I'm saying is that if that doesn't take intelligence, then I don't know what intelligence is."[37]

Haptic intelligence was also present in people, although it was understood to have atrophied as a consequence of living in an industrial-technological society. In her 1974 book *New Mind, New Body,* biofeedback expert Barbara Brown asserted that the sense of touch was uniquely conducive to feelings of "peace and understanding," and that modern Americans critically undervalued it: "The voicings of the skin cut through convoluted language and the pressures of consciousness with a remarkable accuracy about the nature of reality." Ecofeminist Elizabeth Dodson Gray likewise called for a renewed emphasis on

> communicating with the earth through the sensuousness of our bodies. Is it possible that our fully sensuous body is the erotic connection to our world which we have been lacking? We know our living is part of the earth's living systems—that we are rooted in the earth and sensually in dialogue with it. Is it possible that intuitive wholistic awareness has as nerve endings our human skin?[38]

Because animals had a highly developed kinesthetic sensibility, imitating their movements was seen to be one means for restoring the lost proprioceptive faculty in humans and for engaging their alien animal subjectivities in dialogue. This was consistent with ecological thinking's emphasis on an epistemology rooted in emulation and empathy. Adopting animal choreographies was believed to allow a person to get inside the mind of the animal and enter into an empathic relationship with it. Gary Snyder wrote, "You learn animal behavior by becoming an acute observer—by entering the mind—of animals. . . . The miming is a spontaneous expression of the capacity of becoming physically and psychically one with the animal, showing [that] the people know just what the animal does." Paul Shepard in his 1978 book *Thinking Animals* described Native American shamanic dancing as providing a mimetic bridge to animal intelligence. He understood such dances to be "saying . . . 'There is a whole other way of knowing which is the way of the falcon or the bison or the bull. I shall dance it and you will realize that you do know it, that your consciousness and thoughts are like and unlike those of the bison.'"[39]

Different approaches to animal choreographies were explored by composer Alvin Lucier, ethologist John Lilly, and dancer Simone Forti. In his composition *Vespers* of 1968 (named after the common North American vesper bat), Lucier directed blindfolded performers to navigate past obstacles and around the stage using only echolocation. Inspired by Donald Griffin's cognitive ethology, Lucier's score instructed the performers to orient themselves using hand-held echolocation devices that emitted clicks at different speeds, so as to move through the performance space like "non-human migrators" or "slow ceremonial dancers." They were to be transported in their imaginations to "dive with whales, [or] fly with certain nocturnal birds or bats."[40]

John Lilly designed a flotation tank where he would float for hours in seawater in order to gain insight into the lives and minds of the sea mammals he was studying. After he abandoned his dolphin laboratories, he used such flotation exercises as a means for communicating the basics of the "dolphin mind" to people who enrolled in his workshops at the Esalen Institute in Big Sur, California, a center of the New Age Human Potential Movement. The key to understanding dolphin mentality, Lilly maintained, was to appreciate what it meant for them to be air-breathing animals living in the water. A severely ill or injured dolphin had to be supported at the surface by its fellows, or it would drown. "This fact alone makes each dolphin dependent upon all other dolphins. There is a mutual interdependence far greater than it is between human beings since an unconscious dolphin will drown unless brought to the surface and woken by his fellows."[41] Sometimes they carried each other for days.

In Lilly's workshops, participants would emulate this interdependency by stepping naked together into a swimming pool. They would then take turns intentionally hyperventilating until they passed out, at which point the incapacitated member would be buoyed up by the others until he or she regained consciousness. Only in this way, Lilly believed, could humans experience on a profoundly unconscious and physiological level the mental lives of dolphins.[42]

As a result of Lilly's ideas, many New Age thinkers adopted the image of dolphins as a model for humans aspiring to ecological consciousness. Dolphins' constant awareness of their own interdependency seemed to exemplify the quality of mind that humans needed in order to correct the ecological imbalances that they had caused by overemphasizing their individual egos. Joan McIntyre wrote in her book *Mind in the Waters* that the dolphin school was itself a supra-personal entity, like a single organism, through which individual dolphins passed like cells that were periodically regenerated. Renewing itself to continue living for generation after generation, "knit together by the integrated sensing of each member, each member sharing his or her information with the others," she imagined, "the school is an ancient, uniquely supportive culture—a creation greater than the sum of its parts. . . . I envy them. I envy their life and the ease of their connections. . . . This is the mind I have always believed existed somewhere."[43]

In her work as a professional dancer, Simone Forti focused on animal choreographies in order to explore the dimensions of "body intelligence." After attending Reed College in the mid-1950s, Forti studied dance in San Francisco with Anna Halprin, who was a pioneer in applying the notions of kinesthetic awareness and body intelligence in the field of modern dance. In the sixties, Forti began working with a group of dancers including Steve Paxton, Trisha Brown, and Yvonne Rainer. While performing with them in Italy in 1968, she visited the Rome Zoo and began thinking about body intelligence in animal movement, adapting what she had learned from Halprin a decade earlier. "It was the first time in years that I allowed myself to be led by the feedback from my body sensations," she later wrote.[44]

In 1969 Forti premiered *Sleepwalkers*, later renamed *Zoo Mantras*. The dance is based in large part on the movements of a caged polar bear at the Rome Zoo. In an artist's statement, Forti attested to the feelings of empathy that her emulation of the polar bear had evoked in both of them:

> The strongest clue I have to why I've again and again returned to the zoo is the feeling state that I'm sometimes in when I find myself looking eye to eye with a particular animal. Often I find that at that moment there's a flood of warmth that seems to pass between us and that I'm held spellbound.
>
> I once observed the pacing patterns of a female bear. . . . I then tried to mirror her and to learn her dance by copying it as best I could from where I was. . . . But she quickly became aware that I was dancing with her. And ending her practice she came over to me as close as she could, stretching her long neck and sniffing in my direction. I too found myself with chin stretched forwards, nostrils flared, trying to nuzzle across the space of hedge and moat.[45]

Forti's animal choreographies developed out of her professional project to recover a style of movement governed by natural "measure":

a rhythm produced by physical feedback loops involving the body, rather than by any artificially imposed steps or tempo. She believed that this had important cultural implications for restoring humans to a sense of their place in nature. In this connection, her aims as a choreographer intersected with those of experimental composers who were interested in just intonation and harmonic resonance as the means of a musical interface with nature. She first met La Monte Young and Terry Riley at Halprin's dance studio in the late 1950s. In 1961 in New York City, she choreographed a piece to accompany the performance of Young's ambient drone composition *2 Sounds*. Forti's dance featured a simple physical feedback system. As one audience member recalled, "La Monte Young's *2 Sounds* plays as Simone stands in a loop of rope suspended from the ceiling. Another performer turns her around and around, and then releases, allowing the twisted rope to unwind and then wind back up in the other direction."[46]

Forti explained her discovery of natural measure in her *Handbook in Motion*, published in 1974:

> Once, on mescaline, I started weaving the upper half of my body very fast in a figure-eight, supporting and compensating with my lower half. I could do it only if I got the exactly correct dynamic and rhythm going—the exact amount of thrust in one direction that I could then recoil and bring back in the other direction and just keep it weaving back and forth. I couldn't have done one side of the movement without the other side. It was like jumping onto a ride. I've often had that feeling about movement, about rides in movement that you can find with your body formations. They remind me of perpetual-motion machines. I guess twirling is the simplest one.

In the same way, she continued, "It seems to me that when the polar bear swings his head, he is in a dance state. He is in a state of establishing measure, and of communion with the forces of which he is a part."[47]

Such weaving, twirling, and swinging embodied specific manifestations of a larger category of natural movement through space and time that Forti called "flow." She described how, in the hippie subculture that she joined for a year in 1968, "flow" was an important principle governing attitude and behavior:

> "Flow" was central to our system, and faith in the flow of which you were a part and by which you would be sustained. And we ate no meat.

To flow was to relinquish a great deal of control over who and what you were going to come in contact or stay in contact with. You did this by suspending all plans until each moment presented itself with its many facets and its clear indications regarding your own most natural gesture. You would seldom have to interrupt a cycle of involvement because you would never have made plans to do something else. So the merging patterns could live themselves out and the whole system could integrate. One result was a very high incidence of coincidence. I came to think of coincidence as harmonic overtones of occurrence, and as twilight guideposts.

There was almost a discipline of speaking only when necessary, only when the clues weren't there to be read, weren't already speaking.[48]

In the seventies, Forti began collaborating with Charlemagne Palestine. She found his music conducive to her explorations of "dynamology"—her term for how each animal's movements embodied the "structural attitudes" particular to its species. She and Palestine first worked together in 1970, organizing a recital at CalArts given by raga master Pandit Pran Nath. After that they began composing and choreographing collaboratively. Forti studied the movements of rabbits and bears for insights into the mentality of each. "Consider the difference in attitude between moving forward as we do, shifting the body weight from side to side, and moving forward as does the rabbit gathering all his energy at once to spring away from the earth by an act which is instantaneous in its bilateral symmetry," she proposed. From three grizzlies in the Central Park Zoo, she learned "a way of turning: from standing on hind legs reaching the head back and twisting it around, tipping the weight and kind of twirling half way around while falling back on all fours." In 1976 she put many of these movements together into a collaborative work that she called *Planet,* in which a group of forty dancers mimicked lions, elephants, monkeys, bears, lizards, and birds all onstage together.[49]

Contact Improvisation

Although Forti learned her animal choreographies by watching animals intently, kinesthetic intelligence implied privileging touch as an epistemological medium over vision. Could humans physically dance *with* animals? Forti's colleague Steve Paxton invented the dance form known as contact improvisation as a way to explore kinesthetic intelligence through touch. Instead of emulating other animals, the dancer of contact improvisation developed his proprioceptive perceptiveness through intersubjective physical contacts with other humans.[50]

In contact improvisation, each dance is created by two dancers, who stay in physical contact for most of the dance and use their points of contact as the basis for a mutual improvisation that relies on a "sharing [of] information through the body." In this way the form connects the two dancers in a feedback loop of touch, which they maintain by staying "intensely focused in moment-to-moment awareness of change." Trading weight and momentum, they produce a shared center of gravity that each responds to but that neither is fully in control of. Dance historian Cynthia Novack reported that while dancing contact improvisation, she "began to experience periods of an effortless flow of movement, not feeling passive, and yet not feeling actively in control either. The sensation of 'being guided by the point of contact' with my partner fitted the description of 'allowing the dance to happen' to me."[51]

Like the sonic ecologies of experimental composers, which used feedback loops to organize the random indeterminacy of John Cage's radically decentralized musical vision, Paxton's contact improvisation form used kinesthetic feedback to reintegrate ensemble dancing after its radical decentralization during the 1960s. Traditional approaches to dance had vested centralized control in a choreographer who, like the composer in classical music, designed the steps that the performers would have to follow. In the sixties, experimental dancers replaced this approach with an emphasis on decentralized group improvisation. In 1970 the experimental dance group Grand Union, of which Paxton was a member, specialized in improvising without a choreographer; but Paxton found the incoherence that resulted from everyone doing his or her own thing deeply unsatisfying. It "open[ed] . . . possibilities," he said, but "eventually led to isolation" of the individual dancers in their private worlds. As a corrective, he found that the reciprocity of physical contact, when added to such improvisation, could restore a feeling of coherence. It produced what he called the "entrainment" of two dancers, connected at the haptic threshold of perception. As he later explained to an interviewer:

> It's about . . . finding the range of delicacy or impulse anyway in the movement, finding the spatial specialness that your partner's mind has . . . but sensing them at a level that is just at the lowest level of your perceptual concentration, something that is almost so small that it almost is slipping away in terms of fragments of time, fragments of impulses in the body.[52]

As in other ecological systems, in contact improvisation, feedback transformed diversity into the basis for new shared possibilities. One dancer recalled how, in the mid-1970s, there was

> a tremendous tension and excitement about encountering anybody, an anticipation, not knowing what was going to happen—whether you were going to dance slowly, hardly move, do a lot of lifting and falling, or whether it was going to be sensuous or kind of playful or combative.

Two dancers in contact improvisation together create an embodied systemic intelligence. As Novack described it, "Contact improvisation defines the self as . . . the responsive body listening to another responsive body, the two together spontaneously creating a *third force* that directs the dance." If successful, the result was a feeling of interconnectedness that relied on "rapport" and that rewarded the dancers with a feeling of "communion."[53]

Horse Whispering

Many of the same principles that informed contact improvisation also defined the horsemanship techniques that Ray Hunt described in the book *Think Harmony with Horses*. He instructed the rider to relinquish egoistic control of the horse and to develop a kinesthetic sensitivity that enabled him or her to acknowledge the horse's feedback, creating a dyadic system. Hunt wrote, "A make-it-happen attitude can defeat your purpose. Just back off and *ride the horse* and know where every foot is. Know how it felt when it got there. Know what happened to get it there."[54]

Hunt was among the pioneers of a new horse-training technique—now popularly known as "horse whispering"—that was widely adopted in the United States during the 1970s as a result of its spectacular success with "problem" horses. Horse whispering was radical in that it forsook the historically ingrained image of good horsemanship as based on the power to dominate or "break" the incorrigibly wild horse. Horse breaking pitted man against animal in a battle by which each tried to instill lasting fear in the other. Horse whispering to the contrary developed the horse-rider relationship as a friendship rooted in trust and empathy. In *Think Harmony with Horses*, Hunt wrote, "[Your horse is] an individual and this is why I say he's entitled to his thoughts just as you are entitled to yours. . . . I want to respect my horse's thoughts and feelings."[55]

Horse whispering aimed to develop horse and rider into one mind sharing the same idea. Hunt instructed his students on the need to respect the horse's contribution to this shared mind:

> This is real important to me. You can ask the horse to do your thing, but you *ask him*; you offer it to him in a good way. You fix it up and let him find it. You do not make anything happen, no more than you can make a friendship happen. . . . If we don't let the horse think a little bit too we're not being fair.

Hunt emphasized that his clinics were not aimed at training the horse but at training the rider in how to learn *with* the horse, by first listening *to* the horse and then learning *from* her. We are, he explained, "working to recognize what our horse is telling us of how we feel to him."[56]

As in contact improvisation, this communication happened mostly kinesthetically, through "feel." "When the horse moves and you move with him, your idea and his idea become one," Hunt directed. "He isn't dragging you and you aren't pushing him along. . . . [Y]ou're in time with his body and then he can be close to you and understand what you are talking about." Hunt explicitly used dance as a metaphor in describing this interaction.

> I use the symbol of "dancing" a lot because there's a rhythm and there are two of you working together and you're feeling of each other. . . . The music is the life in your body and in the horse's body. . . . Feel it. A feel following a feel—there's no pressure mentally or physically. That's what you are offering him. . . . You reach out, you feel *of* him, then *for* him, and then you both feel together.[57]

An understanding of horse whispering spread beyond the horse-riding community to the general public in the late 1970s, with two successful Hollywood movies popularizing the approach in 1979. *The Black Stallion* was directed by Carroll Ballard, who would later direct *Never Cry Wolf. The Electric Horseman* starred Robert Redford, who was himself an avid horseman. Redford was widely known at the time as both an equestrian and an environmentalist, having written a *National Geographic* story and a coffee-table book about riding horseback through the Western wilderness.[58] Both *The Black Stallion* and *The Electric Horseman* showed people how rewarding it could be to develop a friendship with a horse.

The Black Stallion explicitly juxtaposed the two relationship styles of horse breaking and horse whispering. The film begins at sea, as an eleven-year-old boy travels on an ocean liner on which a "wild" stallion is also being transported. In the opening scene, we witness a violent example of old-style horsemanship, as the horse fights blindfolded against a group of men who use ropes and whips to force him into a stall. After the ship sinks in a storm, however, the boy and horse find themselves alone together on a desert island and develop a symbiotic relationship. The boy determines to ride the horse, and lays the groundwork for this in what the movie self-consciously frames as an archaic ritual. This begins with the boy (now dressed only in a loincloth) leaving an "offering" of food for the horse, and progresses in a slow series of approaches and retreats between the two. The entire ritual unfolds wordlessly, and eventuates in a three-movement dance duet for boy and horse. The first movement ends with the horse eating from the boy's hand (fig. 11). The second movement ends with the boy hugging the horse. And the third movement ends with the boy riding bareback along the beach, his lack of riding gear a symbol of their natural oneness.

Dan "Buck" Brannaman, who was to become nationally famous as *the* horse whisperer, was already beginning to establish his own reputation by this time. In the early 1970s, Brannaman had been a rodeo performer and did television commercials for Sugar Pops cereal, whose ad campaign featured a cowboy who whipped the corn to make it pop. Brannaman was deeply disturbed by the abusive horse-training techniques that he encountered in his career, and began using Ray Hunt's

Figure 11. End of the first movement of the horse-boy ballet from *The Black Stallion.*

methods to provide therapy for abused and troubled horses. He called his technique "dancing from a distance." His description of its dynamics closely matches the structure of the boy-horse dance scene in *The Black Stallion*:

> When the distance between you and the horse becomes comfortable to him, you start to draw him in. You do this by moving away as he begins to acknowledge you. . . . The farther you move from him, the closer he moves to you. This is known as "hooking on," and it's an amazing feeling.[59]

Robert Redford's character in *The Electric Horseman* was given a life story very similar to Buck Brannaman's background. A former rodeo star, Sonny Steele now rides a horse as he shills for a breakfast cereal company. Eventually, however, this brings him into unbearable conflict with his true feelings about his own identity as a cowboy (fig. 12).

The Electric Horseman depicts both Sonny Steele and his horse as victims of a corporate-industrial society. The corporate executives at Ampco literally steal Sonny's identity. He is their "corporate symbol," the "Ampco Cowboy," pimped out in jeweled boots and electric lights. The experience has driven him to drink. In Las Vegas, after he tries to counsel the corporate suits on the care of the horse that he is riding, they simply replace him with someone else whom they now call Sonny Steele; people, they point out, can't even tell the difference. Ampco similarly treats Rising Star, the horse that Sonny rides, as just their piece of property, "a substantial investment" made in order to

Figure 12. Robert Redford as Sonny Steele in *The Electric Horseman*.

"sell products." "Bring in the damn horse," they growl, when it is time for a photo shoot.

Sonny recognizes the horse's exploitation even though he can't see his own: he notes that it's drugged, and that its injured tendon has not been braced and bandaged because it wouldn't look good for the cameras. He alone sees Rising Star as an intelligent and sentient being, with "more soul than most people," and determines to save him by stealing him. In rescuing his horse, he also rescues himself; he regains his self-respect and finds love. With the help of a supporting cast of mostly blue-collar Americans, he evades the official authorities and turns the horse loose in the Rocky Mountains. The emotional high point of the movie comes when Rising Star joyfully unites with a herd of wild mustangs. Sonny and Rising Star have mutually helped each other to find more fulfilling lives.

In both his movie roles and his private life, Redford offered the American public a new variation on the cowboy archetype, one that might be called the "biophilic cowboy" (see plate 6). He used Buck Brannaman's language to describe his relationship to the horse that he rode in the movie (who was aptly named Let's Merge). "I got on him," he told an interviewer, "and we both hooked in." When filming ended, he bought Let's Merge and maintained their relationship for the next eighteen years, until the horse died.[60]

Horse whispering's substitution of a trust-based relationship for the violent heroism of "breaking" the horse entailed the transformation of the cowboy as an American symbol. It implied a revision of the myth of the Old West as a heroic "conquest" of the wilderness and of Native Americans—and a revision of the ideal of American manhood that went with that myth.[61] Bruce Schulman, in his book on American culture and society in the seventies, discusses the era's revision of normative masculinity from the "tough, even violent, and fiercely independent" type to "the sensitive New Age male," characterizing it as a "shift from John Wayne to Alan Alda."[62] But it might be more accurately framed as a shift from John Wayne to Robert Redford.

Choreographies of Confinement

Robert Redford's image as a biophilic cowboy exemplifies both the possibilities and the limitations of affective human-animal relationships. For, even when framed as a kinesthetic dialogue, horse riding remains in many respects an inherently unequal relation, in which the human rider retains most of the control. The same applies to the relations between dolphins and their trainers. Friendships between humans and domesticated animals occur within a structural inequality of power relations. Thus while some ecological thinkers saw these relationships as offering opportunities for mutual enrichment, others defined domestication of any sort as a diminution of the animal's personhood.

In the context of a discourse privileging animals' haptic intelligence, their domestication was understood to entail a loss of the unique skill set that they had evolved naturally as a species. Paul Shepard wrote in his 1978 book *Thinking Animals*, "The mushy aspect of kept animals is their loss of species form, that firm outline which explicitly associates them with niche and kind. . . . Cut from the larger ecosystem, kept animals become pseudopeople who enter the household trailing unraveled ends of a tattered integrity." There was a disappointing limit, he concluded, to what could be learned from the domesticated animal. It was "no longer a medium by which contemplation of the nonhuman is the access to the equipoise between unity and diversity, nature and culture, self and Other."[63]

To thinkers like Shepard, the practice of detaching animals from their natural environments and confining them in proximity to humans was an infliction of injury, and therefore a form of cruelty. Jane Goodall observed that "the zoo chimp has none of the calm dignity, the serenity of gaze, or the purposeful individuality of his wild counterpart." Gregory Bateson averred that dolphins and whales who were kept in tanks were "bored to tears" most of the time. John Lilly more pointedly compared his former dolphin laboratory to "a concentration camp for my friends."[64]

In his documentary films *Primate* (1974) and *Meat* (1976), Frederick Wiseman used cinematographic techniques to emphasize how a technology of cages, pens, and chutes was employed in behaviorist laboratories and slaughterhouses alike to restrict animals' movements and produce a choreography of slow violence.[65] This choreography produces insistent, unidirectional, horizontal flows that in both films are shown to channel the animals inexorably toward death and dismemberment. *Meat* begins with a shot of wild buffalo wandering on the plains; from there it cuts to cattle on the range, and then to cattle in feed pens. Via a series of chutes and gates through which they are forced by men, the penned cattle are by stages transformed into beef carcasses that progress down a disassembly line, ultimately becoming a flow of shapeless "ground beef" extruded from metal grinders. In *Primate*, the camera similarly tracks the forcible transport of apes and monkeys from their stationary cages to other, rolling cages, to pieces of experimental apparatus, and finally to the dissecting table.

Like *The Electric Horseman,* ecological texts drew parallels between the plight of animals whose movements were forcibly restricted and the lot of human beings trapped in a dehumanizing social system. Wiseman's *Meat,* for example, divides its time between the slaughterhouse's disassembly line and a chronicle of labor disputes between management and workers. In one scene, the workers and carcasses, both draped in white plastic, are difficult to distinguish from one another. Godfrey Reggio's *Koyaanisqatsi* depicts the life of humans in industrial society as a compulsively linear choreography similar to those in Wiseman's films, which exists in marked contrast to the swirling forms of nature. As film critic Gregory Stephens observed, "The visual narrative reiterates the motif of channeled motion: people and the objects they create move along channels which seem to follow the same logic: conveyer belts, freeways, etc." In one scene, the editing cuts back and forth between two shots: a machine spews out hot dogs, and people riding escalators on the subway emerge in homologous long lines. The industrial system, it implies, offers two varieties of processed meat (fig. 13).[66]

Empathy for the human animal was often present as a theme whenever a text or practice focused on promoting empathy for animals confined by the choreographies of modern civilization. It appeared both early and late in the decade, in films as diverse as *Meat* and *The Electric Cowboy, Koyaanisqatsi* and *Rodeo.* Carroll Ballard's twenty-minute documentary *Rodeo* was made in 1969, before he was chosen to direct *The Black Stallion.* Like Reggio's *Koyaanisqatsi,* it is experimental in form. Filmed at the 1968 National Finals in Oklahoma City, *Rodeo* was made on a $20,000 budget provided to Ballard by Marlboro cigarettes, which was promoting the rodeo cowboy as an image of the "Marlboro man."[67] But Ballard's cinematography undercuts the mythical status of the rodeo cowboy by redirecting the viewer's attention away from the artificial cheer of the stands, to focus behind the scenes on the slow violence of pens and prods and noise that would otherwise escape notice. In this way the film ironizes and subverts the rodeo's traditional associations, to offer a portrayal of the cowboy and the animals he rides as equally the victims of the social system.

The plot of *Rodeo* is simple: a cowboy in a red shirt tries to ride a bucking bull, but gets thrown off and scores a zero. He and the other cowboys remain casually stoic. This affective conditioning is understood to be a requirement of the position. Their physical and emotional suffering must remain for the most part inaudible and invisible,

Figure 13. Two varieties of processed meat, as depicted in *Koyaanisqatsi.*

although at one point we glimpse a man limping out of the arena. However, Ballard offers many extreme close-ups of both bulls and cowboys; the camera scrutinizes their eyes, the cowboys' hands, and the bulls' wet and vulnerable snouts. We see both antagonists in the bull-riding drama against a common backdrop of bars, implying that the cowboys, too, are in a zoo of sorts (see plate 7).

In this shared context, the bulls are shown to "speak" for the cowboys; they express the fear and rage that the cowboys are prohibited from displaying. As the two species engage in the prescribed choreography, they dance a perverse variant of contact improvisation in which every point of contact is a source of pain. Ballard's cinematography supplies the empathy that is missing in the rodeo itself, in order to show that it is lacking. As the red-shirted cowboy tries to ride his bull, slow-motion close-ups bring into focus the spit and snot streaming from the bull's nose. When the cowboy is thrown into the dirt, Ballard superimposes the sound of the bawling bull over a close-up of the cowboy's open mouth: the cry could be coming from him (see plate 8).

A Tale of Two Dolphins

The choreographies of confinement in Robert Redford's *The Electric Horseman* are disrupted when Sonny Steele absconds with Rising Star and releases him into the wild. In reality, the practice of "liberating" captive animals from their tanks and cages was one that by the end of the decade was bringing out some of the differences in outlook within the community of ecological thinkers.[68] The logic of the ecological worldview as it applied to animals led some of its most faithful adherents to adopt radical tactics, called by some "ecoterrorism." Paul Watson, for example, having been voted out of Greenpeace for seeking violent confrontations with Canadian seal hunters, in 1978 formed the Sea Shepherd organization, which engaged in "direct action" to save the whales by ramming a whaling ship in 1979.[69]

The question of animal liberation pitted radical environmentalists against their more moderate brothers in arms. Despite the increasing influence of a discourse equating captivity with cruelty, the romantic fantasy of returning captive animals to the wild was problematic in practice. Notably, after filming *The Electric Horseman*, Robert Redford did not free the horse Let's Merge somewhere in the Rocky Mountains, as his character Sonny Steele did with Rising Star in the movie. There remained considerable uncertainty as to whether even animals born in the wild, once domesticated, could reestablish a place for themselves if released back into the wild. Dolphins and horses might need to situate themselves within familiar schools and herds in order to survive.[70]

A case that unfolded in Hawaii at the end of the decade exemplified the divergent positions among ecological thinkers concerning the ethics of keeping animals in captivity versus liberating them. In May 1977 Kenny LeVasseur and Steve Sipman, two twenty-six-year-olds recently graduated from the University of Hawaii, took two bottlenose dolphins named Puka and Kea from the university's Kewalo Basin Marine Mammal Laboratory and released them into the wild. They left in their place two plastic porpoises on which they had written graffiti messages: "Let my people go" and "Slaves no more." As they later explained to the press and the authorities, their "Undersea Railroad" action was based on a "moral and philosophical commitment to the idea that man has no right to capture, or hold in captivity, intelligent, feeling beings."[71]

The dolphins legally belonged to Lou Herman, a cognitive ethologist who was using them in a research study to ascertain the extent of dolphins' ability to grasp human language. He had been teaching Puka and Kea the basic idea of a language built from a logical string of signifiers. Kenny and Steve were his lab assistant's assistants, responsible for the dolphins' day-to-day care. But Steve, like John Lilly, had come to believe that the dolphins' intelligence put them "above science" and that their behavior should only be studied "in the field" in a natural context. He claimed that he had meditated by Puka's tank and that she had told him telepathically that she wanted to be free. Kenny also thought that the dolphins were aware of their captivity and miserable about it.[72]

Steve and Kenny believed that Herman's experimental regimen verged on cruelty. He did not allow them to feed the dolphins as much food as they needed. He used the "slow, precise, repetitive routines" of operant conditioning, which they felt Kea and Puka found "tedious" and "demoralizing." But Ken Norris, a dolphin trainer and biologist at the University of California, Santa Cruz, explained to a journalist that the nature of Herman's research into the limits of dolphin intelligence required that. "Herman is a terribly precise worker," he said. "The animal has to go through a series of repeated tasks that are refined more and more until you reach a threshold. The animal will be frustrated, because you're testing its limits." He implied that Kenny and Steve were slackers, unable to appreciate the rigor of the experimental process: "To someone who doesn't have any discipline himself, it may seem kind of harsh."[73]

Kea and Puka, at any rate, seemed to feel differently about their training regimen. A major source of conflict between Herman and the two animal liberationists was Puka's beloved Frisbee. It was her favorite toy; she could catch it and had even learned to toss it back by spinning it on her snout. But while Kea demonstrated her eagerness to start the daily regimen by pushing all her toys to one corner of her tank to be removed, Puka would show her reluctance by taking the Frisbee with her to the center of her tank and sinking down to the bottom with it. The struggle of getting it back was impeding the training sessions—once, Puka's entire tank had to be drained before the toy could be gotten away from her—and she was never in the right frame of mind to begin her training afterward. Herman therefore decreed that Puka could only have the toy if it was secured to a rope for easy retrieval. But, in what must have seemed like an apt metaphor, tying a rope to the Frisbee made it unable to fly and ruined Puka's interest in it as a toy. So while Kenny and Steve publicly conformed to Herman's edict, they surreptitiously kept cutting the Frisbee's tether. Herman then declared that Puka should thenceforth have no Frisbee, period. Fed up with what he saw as LeVasseur's and Sipman's rebellious juvenile antics and irresponsible disregard for his scientific work, Herman had them fired. That was on the Friday before Memorial Day weekend. Before dawn on Monday, the two transported both dolphins by van to the ocean.[74]

When news of the dolphins' release made the papers, public opinion was at first inclined to favor Kenny and Steve; but it shifted as Herman and Norris succeeded in portraying them as hippies who didn't really know much about dolphins. Kea and Puka were Atlantic bottlenose dolphins, Herman informed the press; they wouldn't find other animals of their own species in the Pacific Ocean. They had been released in shark-infested waters. Freeing them, he asserted, was tantamount to killing them. "It's a watery jungle out there," he insisted. "It's out there in the ocean that everything bad is happening—that porpoises are being killed by tuna fishermen."[75] The meaningful protection of dolphins and other cetaceans depended on the kind of scientific work that he and his colleagues engaged in. To avoid being cast as the coldhearted villain in the drama, Herman hastened to inform the press that he, too, had an empathic relationship with his dolphins. Kea liked learning, he reported, and whenever she made a breakthrough, she expressed delight by cavorting around her tank, throwing the fish that she got as a reward up in the air and catching it. Turning the tables on the accusation that he was a cruel experimenter, he told one reporter, "To take an animal and toss it into strange waters, ask it to socialize with animals, hunt when it had been fed, ward off predators, etc., is the cruelest experiment anyone could have done." Over the next two days, Kea was spotted twice along the beach. One of her eyes was swollen nearly shut, suggesting that she had wounded herself by scraping up against a coral reef. She seemed to be in distress, yet four times she eluded recapture. Then she disappeared. Herman told a journalist, "I had been with Kea for eight years. When I saw her thrashing about in the water at Makua—to see her alone in such shape and so frightened in these waters where these guys had dumped her . . ." He became too distraught to finish his sentence.[76]

Was liberating Puka and Kea an act of civil disobedience akin to the Underground Railroad's courageous defiance of laws supporting human slavery? Or was abandoning them to fend for themselves a romantic evasion of moral responsibility? Journalist Arthur Lubow wrote an extensive treatment of the controversy for the general-interest magazine *New Times*, attempting to sort out the ethical issues involved. While acknowledging the shortcomings of the two animal liberators (Kenny, he suggested, was a bit of a blowhard; and Steve, a bit of a space cadet) he ultimately sided with their viewpoint. As he reasoned,

> Dolphins and apes seem capable of abstract thoughts. They are aware of the confines of their prisons. Unhindered by man, they would lead lives of hunting, mating, playing. In tanks or cages, they conduct a crippled existence. For some animals—and while the exact boundary is debatable, it must lie somewhere above rats and somewhere below dolphins—the loss of freedom is perceived as suffering. If these animals have rights they must be released. . . .
>
> When Lou Herman defends keeping dolphins in captivity, he refers to his concern for their well-being, the benefits to more highly developed humans from this work, the dolphins' evident contentment with their captive state, the burden of a painful history, and the ultimate importance of his research to human civilization. All of those arguments have been used before. By slaveholders . . .
>
> Herman's is the voice of intelligence and civilization. Yet listening to him, you hear echoes of the old rationalizations. It must have been similar talking to Thomas Jefferson or John Calhoun, both men of enormous learning and perspicacity, both fully capable of rationalizing human slavery: Yes, there are awful crimes committed at other plantations, but here the well-being of the slaves is always the paramount concern—and really, with these lovable but simple-minded creatures, what is the alternative to slavery? And what about the glories of Southern civilization?

Dolphins, Lubow concluded, were similar enough to people that their captivity was a form of slavery. Even if Puka and Kea did not survive long in the open ocean—and this assertion of Herman's was debatable, as there were those who claimed to have spotted them months afterward in the company of other dolphins from related species—who could say with authority that they were better off in Herman's shallow tanks?[77]

Such considerations did not carry the day in court. Charged with the theft of government property, Sipman and LeVasseur tried unsuccessfully to invoke the Thirteenth Amendment banning slavery and involuntary servitude; but the judge rejected all such arguments out of hand, reiterating that legally Puka and Kea were property, not persons. The defendants were convicted.

Nevertheless, the difficulty of distinguishing clearly between the dolphins as property and the dolphins as persons continued to plague the trial in the later sentencing phase. Both defendants were given community service in lieu of six months in jail. But Sipman was also ordered to reimburse the state for the value of the animals. And there a thorny problem arose—for it was hard to set a price on them.[78] Their intelligence, their education, their familiarity with what their keepers wanted of them—all the characteristics that were due to the very sentience that their legal status as property denied—were ironically the primary sources of their value to the state. The ambiguous morality of their confinement and release could not be laid to rest.

From "dynamology" to contact improvisation; from echolocation to flotation therapy; and from horse whispering to animal liberation, the culture of feedback developed a wide range of new practices grounded in the belief that human-animal interactions needed to be intersubjective relationships. This imperative was rooted in the conviction, supported by general systems theory, that animals had minds and feelings that were inseparable from their physicality and were accessible to humans only by means of such intersubjective "dances" and not via the traditional scientific method. In consequence, dancing with animals became part of the cultural repertoire—along with attending to the ambient soundscape, talking with plants, and emulating traditional Native Americans—by which adherents of this rapidly expanding counterculture attempted to resituate themselves in nature's feedback loops and, by so doing, evolve.

SEVEN

Neo-Orthodoxies

Greed is all right . . . greed is healthy. IVAN BOESKY, 1985

Toward the end of the seventies, the culture of feedback came under attack by conservative critics who challenged some of the mainstays of ecological thinking. Molecular biologists reasserted a reductionist approach to descriptions of natural selection, dismissing the theory of coevolution as unscientific. These neo-Darwinists relied on game theory to argue that even altruistic behaviors could be fully explained as the result of competition among "selfish genes."

Similarly, conservative economists applied game theory to the problem of "stagflation"—the late seventies phenomenon of accelerating inflation and rising unemployment—and arrived at a neoliberal solution. Their supply-side economics recommended an end to government regulations in order to "free" the market. Contrary to ecological thinkers like Kenneth Boulding, these neoconservative economists positioned the competitive marketplace, rather than the ecosystem, as the final arbiter of human destiny. Although Norbert Wiener had warned in the second edition of *Cybernetics* against the "simple-minded theory" that the market was a homeostatic mechanism for maintaining social health rather than a "game," economists of the eighties imagined it as both.[1]

Supported by supply-side economic theory, the cultural politics of Reaganism mobilized a popular blend of individualism and optimism to flout ecology's accepted truth that there were natural limits to economic growth. Consonant with a regressive notion of individualism, Reaganism emphasized closing borders and strengthening barriers in order to protect a traditional American identity. This ran counter to ecological thinking's core belief that identity was emergent from the complex dynamics of interdependency characteristic of nested open systems.

Genetic Determinism

In the late seventies and early eighties, conservative scientists attacked the theory of coevolution and rejected the ecological vision of a post-Cartesian science. British geneticist Richard Dawkins was among those who led the charge. He ridiculed the Gaia hypothesis as "wishy-washy" wishful thinking and an "overrated romantic fantasy" invented by people searching for a caring mother in nature. The rain forest, he wrote, was not an intelligence woven of "networking" and "mutual dependence," but an aggregate of disparate self-seeking entities: "an anarchistic federation of selfish genes, each selected as being good at surviving within its own gene pool." In the pages of *CoEvolution Quarterly* in 1981, American molecular biologist Ford Doolittle echoed Dawkins's critique.[2] Defending the theory of coevolution, Lynn Margulis went on the counterattack, alleging that molecular biologists like Dawkins were apparently "incredibly ignorant" of many aspects of biology, including cellular structure, botany, and biochemistry. These subfields, she insisted, lent support to the coevolutionary hypothesis. Meanwhile Dawkins's genetic theories, ascribing agency to individual genes, were as "ridiculous" as the discarded pseudoscience of phrenology.[3]

The primary point of contention in this acrimonious debate was the question of what constituted the "unit of selection" in natural selection. Do populations acting as a group compete for survival? Or does each individual organism compete on its own? Ecological thinkers believed that multiple units of selection were simultaneously in operation, consistent with their vision of nature as a multi-leveled structure of nested open systems. James Lovelock wrote that natural selection reflected the "tendency in this Universe for *complex systems* to be stable and survive." Margulis defended her view that evolution was a process of "symbiogenesis" involving "mergers" of multiple simpler structures into a single entity.[4] But a group of biologists led by Dawkins, John Maynard Smith, and William Donald Hamilton argued that all of evolution could be explained by selection on the level of the individual genome alone; and that therefore, it was unparsimonious and unscientific to posit additional units of selection.

The reason why this debate raged in the front pages of American general-interest magazines like the *New York Review of Books,* rather than in the obscure pages of scientific journals, concerned its relation to Harvard biologist E. O. Wilson's attempt at a grand evolutionary synthesis in his 1975 book, *Sociobiology.* Wilson envisioned the possibility of scientifically explaining how natural selection linked together factors of genetic inheritance, environmental pressures, and social behaviors. Social and physiological systems, he wrote, were "self-regulated by internal feedback loops." He believed in group selection. Yet—for reasons that later became difficult to ascertain—in his attempt at a synthesis, he did not maintain this holistic viewpoint. Instead of describing cultures and genomes as mutually constraining subsystems, he emphasized only how "genes hold culture on a leash." In the introductory chapter of *Sociobiology,* he advised his readers that his ultimate aim was to "biologicize" the social sciences, in order to be able to "predict features of social organization" (including "dominance hierarchies, role differentiation" and even sexual moralities) from a knowledge of a population's genotype and phenotype.[5]

Wilson's *Sociobiology* was published at an intensely political moment in the history of science. It therefore reverberated like the sound of a falling rock that, mistaken by two waiting armies for a gunshot, precipitates repeated volleys from both sides. In both America and Europe by the mid-1970s, the influence of Marxism in the academy was being challenged by conservative scientists who accused Marxists of having politicized intellectual life and created ideologically motivated barriers to the disinterested pursuit of knowledge. At the same time, Marxism's influence among both students and faculty, especially at elite institutions like Harvard and Berkeley, remained strong. Marxists saw the individual as a product of social forces. By advocating an approach that treated all social institutions as epiphenomena of genetically determined individual behaviors, Wilson seemed to be siding with conservatives.

The coevolutionary idea of group selection was consistent with the socialist perspective; its conservative critics, conversely, favored the theory of individual selection. In 1962 British biologist Vero Wynne-Edwards had published *Animal Dispersion in Relation to Social Behaviour,* offering what he believed was scientific evidence for the theory of group selection. He analyzed several case studies in order to demonstrate that animal societies achieved an evolutionary advantage by adopting social conventions that put constraints on selfish individual behaviors. From a political point of view, the implications of this argument favored increased social regulation, such as would be implemented over the next two decades by Harold Wilson's Labour Party in Britain and by environmentalists in the United States. However, Wynne-Edwards's conclusions were quickly set upon by a group of conservative scientists. John Maynard Smith began questioning them in print in 1964. American biologist George C. Williams came to Maynard Smith's aid with a book titled *Adaptation and Natural Selection*, published in 1966, in which he explained away all of Wynne-Edwards's examples of group selection by providing plausible hypotheses for each based on individual selection alone. In a 1973 address, William Hamilton claimed that "Marxism, trade unionism, and fears of 'social darwinism'" were alone responsible for the widespread belief in group selection among biologists, when it was not at all needed in order to explain social behaviors.[6]

The most powerful weapon in the arsenal of the neo-Darwinists proved to be Hamilton's mathematical equations, first published in 1972. Hamilton used decision models taken from game theory to demonstrate that there was no need to posit group selection to explain altruistic behaviors. Hamilton's equations demonstrated that what motivated altruistic behavior could be not group selection, but "kin selection." This made altruism, otherwise a sticking point for the orthodox Darwinian model, compatible with the claim that the only unit of selection was the individual genome. Because genetically one "is" one's genome, an organism could in effect select "itself" for survival (the passing on of its genome) by sacrificing its life to preserve the lives of a sufficient number of related genomes. One genetic determinist famously joked that he would lay down his life for two brothers, or eight cousins. Working out this calculation for various more complex scenarios was where the equations of game theory proved useful. All that was necessary was to replace the mythical "rational actor" of game theory, who sought to maximize his competitive advantage over rivals, with the abstraction of an ostensibly genetically programmed behavior seeking to emerge as the "evolutionarily stable strategy." Relying on Hamilton's equations, Maynard Smith in his 1972 book *On Evolution* introduced the now famous "hawks" and "doves" game scenario, which pitted two such behaviors against each other in different social contexts. Hawks would escalate a confrontation and fight until impaired; doves would display aggression but retreat if hostilities escalated. Depending on the initial conditions, different ratios of each type produced an "evolutionarily stable" population. With the help of

Hamilton's equations, Maynard Smith wrung a full public recantation from Wynne-Edwards in 1975—the very year that Wilson's *Sociobiology* was published.[7] Many of Wilson's readers were filled with dismay to find that he had relied on Hamilton's equations and adopted the theory of kin selection in his grand synthesis.

Wilson may simply have been jumping on the bandwagon without fully realizing its political implications. Some of the other language in *Sociobiology* showed the influence of holistic thinking. "The higher properties of life are emergent," he wrote in its early pages; and "the recognition and study of emergent properties is holism," which he opposed to the "reductionism of molecular biology." His oft-disparaged statement that "genes hold culture on a leash," which appeared in his 1978 defense of *Sociobiology* titled *On Human Nature*, was typically quoted out of context by his critics. In context, its thrust is not to deny agency to social institutions, but to reject a Cartesian dualism in which culture could be imagined to float free of material constraints—hardly a position that a Marxist could disagree with. "Can the cultural evolution of higher ethical values gain a direction and momentum of its own and completely replace genetic evolution?" Wilson asked.

> I think not. The genes hold culture on a leash. The leash is very long, but inevitably values will be constrained in accordance with their effects on the human gene pool. The brain is a product of evolution. Human behavior—like the deepest capacities for emotional response which drive and guide it—is the circuitous technique by which human genetic material has been and will be kept intact. Morality has no other demonstrable ultimate function.[8]

Read in full, this passage is not even clearly reductionist. And yet it is not quite right, either. The "leash" metaphor is a disturbing choice, implying that genes are the "master." Why not say "tether" instead? And there is a brazenly missed opportunity, given the fallout that had already materialized from his 1975 publication, to emphasize that it is culture, according to the theory of epigenetics, that holds genes on a leash as well.

Perhaps the most surprising thing about the reaction to *Sociobiology* in retrospect is how genuinely surprised Wilson seems to have been by it. Moderate criticisms were quickly overwhelmed by a tide of vehement recriminations. Epigeneticist Conrad Waddington wrote a temperately critical review for the *New York Review of Books*, faulting Wilson for slighting "the reciprocal interaction" between behaviors and gene selection and for including "no mention of mind" as a phenomenon of complex systems. However, Wilson's leftist colleagues on the biology faculty at Harvard, having gotten wind of the moderate tone of Waddington's review, joined a dozen other signatories in a public letter to the *Review*, in which they forcefully accused Wilson of allying himself with eugenics, militaristic nationalisms, and sexism.

Anthropologist and antiwar activist Marshall Sahlins rushed to print with a book titled *The Use and Abuse of Biology*, in which he called Wilson to account for his uncritical adoption of game theory. "The theory of sociobiology has an intrinsic ideological dimension, in fact a profound historical relation to Western competitive capitalism," Sahlins wrote. Via game theory's equations, natural selection was recast as the "optimization or maximization of individual genotypes" through "the exploitation of one organism by another." The excessive emphasis that game theory placed on individual decision making, Sahlins continued, ignored the truth that social systems produce their own dynamics: "What is reproduced in human cultural orders is not human beings qua human beings but the *system* of social groups, categories, and relations in which they live," Sahlins argued. "Human reproduction is engaged as the *means* for the persistence of cooperative social orders."[9]

At the 1976 meeting of the American Anthropological Association, a formal motion to censure Wilson and his sociobiology as reductionist and racist was narrowly defeated only when the grand dame of cultural anthropology, Margaret Mead, literally rose (supported by a cane) to defend Wilson's academic freedom. Wilson fared less well at the American Association for the Advancement of Science conference in 1978, where when he rose to speak, he was met by student leftists belonging to the International Committee Against Racism who chanted, "Racist Wilson you can't hide, we charge you with genocide!" before running up to pour a pitcher of ice water on his head and yelling, "Wilson, you're all wet!"[10]

While Wilson cried foul and accused his critics of "academic vigilantism" and ideological bias, Richard Dawkins doubled down on the genetic determinism in Wilson's argument. All that was needed in order to explain altruism, he argued, was an understanding that the individual being was merely a pawn of its own "selfish genes." He wrote in his influential 1976 popularization of kin selection, *The Selfish Gene*, "We, and all other animals, are machines created by our genes." "Like successful Chicago gangsters," our genes find ways to survive "in a highly competitive world"; as a result, a "predominant quality to be expected in a successful gene is ruthless selfishness."[11]

Dawkins did not even pay lip service to the coevolutionary theory that a web of mutual causality linked genetics and behavior, as Wilson had. The "causal arrow," he insisted, pointed only from gene to behavior. In the face of his own materialistic determinism, he found solace in a return to the Cartesian dualism. Consciousness, he believed, enabled humans uniquely among all living species to transcend their genetic predispositions. Humanity "alone on earth," he wrote, had a "capacity for conscious foresight," which enabled it to "rebel against the tyranny" of its selfish genes.[12]

Stephen Jay Gould took up the gauntlet that Dawkins had thrown down. Over the next decade and more, the two engaged in a well-publicized debate over the mechanism of evolution, the most salient part of which concerned the question of whether evolution proceeded uniformly, or in a "punctuated stasis" characterized by long periods of stability interrupted by intervals of rapid change. The former hypothesis was consistent with Dawkins's contention that randomly mutating genes were the only unit of selection, since the probability of the occurrence of beneficial random mutations should be constant over long spans of time. Gould's "punctuated equilibrium," on the other hand, implied a homeostasis produced by complex patterns with multiple levels of selection, which would at rare intervals negotiate a coevolutionary turning or tipping point. In setting forth his argument, Gould cited both Gregory Bateson and Arthur Koestler. Those thinkers, he wrote, offered an alternative model to genetic reductionism in their vision of nature as a nested hierarchy of semi-autonomous systems.[13]

Late in his life (he died in 1980), Bateson also published a final book in which he tried once more to make the case for ecological thinking in the face of this new challenge. In *Mind and Nature: A Necessary Unity*, published in 1979, Bateson defended a holistic approach to human nature, as against Dawkins's dualism of genetic determinism and a transcendent consciousness. Reiterating his rejection of Occam's razor as a sound scientific principle, Bateson insisted that simplicity was not a test of truth; nor did the existence of a simpler explanation constitute a refutation of a more complex one. Natural selection operates on every level of the multitudinous hierarchy of nested systems into which nature organizes itself. The multiplicity of semi-autonomous levels, he explained, and not a Cartesian transcendence, was responsible for the high degree of freedom-to-change that was observable in phenomena resulting from complex systems—like human consciousness or evolution.[14]

In the end, the difference between the supporters of coevolution and the proponents of neo-Darwinism proved less marked in their scientific positions than in the worldviews that they stood for. Gould was an ecological thinker influenced by Koestler and Bateson. Dawkins was a crusader for the independence of science from outside influences, among which he numbered politics and spiritualism.[15] In that sense, Dawkins was part of a broader pushback on the part of traditionally minded scientists against ecological thinking and its call for a move "beyond reductionism" toward a post-Cartesian metaphysics. Becoming more vocal in the second half of the seventies, those scientists, including Carl Sagan and B. F. Skinner, defended the scientific project as a quest for facts about a world of objects, and denied the legitimacy of telepathy and other paranormal phenomena that attributed mind to nature.[16]

Unintended by any faction in the scientific community, however, was the way in which their public debate over the merits of competing evolutionary theories provided grist for creationist arguments made by a neoconservative religious Right that was rapidly gaining political influence. The Christian Right shared conventional scientists' dismay at the rise of such belief systems as neo-paganism and ecofeminism. But they were also adroit at turning the scientists' own words against them in order to discredit evolutionary arguments generally. "Harvard Scientists Agree Evolution Is a Hoax" claimed the title of one creationist pamphlet, citing Gould's refutation of Dawkins's arguments. During the 1980 presidential campaign, Ronald Reagan publicly questioned the validity of evolutionary theory while speaking before a group of Christian fundamentalists in Dallas, maintaining, "It is a theory, it is a scientific theory only. And in recent years it has been challenged in the world of science and is not believed in the scientific community to be as infallible as it once was. I think that recent discoveries down through the years have pointed up great flaws in it." If the theory of evolution was taught in public schools, he assured his audience, so should the alternative theory of creationism be taught in them as well.[17]

The Resurgence of Game Theory and the End of Limits

In the early eighties, the public prestige of game theory was on the rise again, not only as it applied to the survival of the fittest in the neo-Darwinian model of evolution, but also in economics, as supply-side economics impacted the politics surrounding government regulation and taxation. According to economic historian Philip Mirowski, 1980 marks the beginning of a "vast outpouring of literature on games."

This literature provided the intellectual justification for the rightward swing of American economic policy associated with neoliberalism. Game theory was influential in the thinking of Milton Friedman, whose monetarist policies were adopted in 1979 by Paul Volcker, chairman of the Federal Reserve, when he decided to shock the American economy out of a period of accelerating inflation by drastically raising interest rates, driving unemployment up beyond ten percent.[18] Economist Lester Thurow in 1980 published *The Zero-Sum Society*, referring to the social system as a von Neumann zero-sum game. In 1981 the book became a paperback best seller.

Thurow used game theory to advocate the deregulation of the transportation and energy industries (on the basis that more aggressive competition in those key sectors would lead everywhere to lower prices) and the elimination of the corporate income tax (on the basis that higher corporate profits would attract investments, which would be plowed back into the businesses to improve their efficiency). Such ideas fostered what came to be known as "supply-side economics." Supply-siders argued that tax cuts would lead paradoxically to an increase in government revenue since they stimulated stronger economic growth. This was the thinking behind the regressive tax cuts signed into law by President Reagan in August 1981. In 1982 economist Vernon Smith began using computer programs to test the outcomes of multivariable game scenarios and announced that according to his results, free-market relations coupled with strict corporate secrecy "together . . . are sufficient to produce competitive market outcomes at or near 100% efficiency."[19] As Mirowski observed, nowhere in this wholesale embrace of game theory did conservative economists acknowledge the poor fit between game scenarios and real life.

The New Right embraced a rhetoric of optimism and unlimited human potential as a political strategy that it wielded effectively to roll back the environmental regulations of the seventies by characterizing environmentalists as prophets of doom and gloom. In 1982, for example, free-market advocate William Tucker published *Progress and Privilege: America in the Age of Environmentalism*, in which he laid claim to an outlook of "optimism" and "progress," in contrast to the "despairing vision" of environmentalists. Reagan's vice president George Bush stated in 1983, "We all have a choice to make: It is between the shrinking vision of America held by the pessimists or the expansive vision—the expansive reality—we are building right now."[20]

The environmentalist emphasis on population control offered a particular target for the New Right's accusation that environmentalists falsified their outlook with an excess of pessimism. As opposed to what Paul Ehrlich had claimed in his 1968 best seller *The Population Bomb*, the free market, Tucker wrote, would transform overpopulation into a variety of economic incentives for "improving our technology for utilizing resources." As an example, he pointed to the developing technology of genetically modified foods. Conservative economist Julian Simon asserted that there was no reason why Earth should not be able to support a population of thirty billion people. In 1980 Simon publicly challenged Ehrlich to a bet in the pages of *Social Science Quarterly*. Simon bet that over the next decade, the price of five strategic metals (chromium, copper, nickel, tin, and tungsten) on the world market would go down. He framed the bet (which Ehrlich lost) as a test of their two opposing philosophies: the environmentalist's narrative of increasing scarcity versus neoconservatives' belief in the power of the free market to drive perpetual prosperity through technological innovation.

The fight against birth control and family planning became a point of cohesion for free-market enthusiasts, social conservatives, and religious fundamentalists. In his 1981 book, *The Ultimate Resource*, Simon called for more people, more cities, and more children, writing, "I believe that this particular value is in the best spirit of Judeo-Christian culture, which is the foundation for much of our modern Western morality. In Biblical terms, Be fruitful and multiply." And he continued, "The standard of living has risen along with the size of the world's population since the beginning of recorded time. . . . Contrary to common rhetoric, there are no meaningful limits to the continuation of this process."[21]

That there were, in fact, no limits to growth and that environmentalists had exaggerated the need for concern was a common theme among neoconservatives. Simon's public bet with Ehrlich brought him to the attention of game theorist and futurist Herman Kahn, who had left the RAND Corporation in 1961 to help found the Hudson Institute, a conservative think tank with ties to the Stanford Research Institute. Kahn enlisted Simon to help him compile a response to the Carter administration's *Global 2000* report. That report, published in 1980 by a task force appointed by the outgoing president, predicted that by the year 2000 the environment would be even more stressed, and people's wealth and well-being even more precarious, than they were in 1980. In *The Resourceful Earth: A Response to Global 2000*, Simon denounced the report as "totally wrong." He pointed out that none of the dire things that environmentalists had been predicting for the last two decades had ever come to pass: mass extinctions due to deforestation, dead oceans, a lack of arable land, a scarcity of water, climate change, or energy shortages. In addition, he reported, the threats posed by industrial air and water pollution "have been vastly overblown." In keeping with his idea that the market could supply whatever was needed to maintain an ever-increasing standard of living, Simon rejected the environmentalist emphasis on sustainability and substituted for it the alternative value of resilience. "Sometimes temporary large-scale problems arise," he wrote; but "the resilience in a well-functioning economic and social system enable[s] us to overcome such problems, and the solutions usually leave us better off than if the problem had never arisen; that is the great lesson to be learned from human history."[22]

Protective Barriers

Game theory's undue emphasis on explicit individual choices, as Steve Heims has observed, generates a "tendency to reduce social and political issues to individual psychology." As applied in both evolutionary biology and economics, neoliberal game theory treated social dynamics as the sum of interactions among independently functioning players, rather than as a macrodeterministic constraint. In his popular 1984 exposition of game theory, *The Evolution of Cooperation*, Robert Axelford identified as the purpose of his book: "to investigate how individuals pursuing their own interests will act, followed by an analysis of what effects this will have for the system as a whole." This is different from how causality is understood in ecological thinking, where, as Manfred Drack has written, "the reaction of the whole is not a result of the reaction of the parts: it is the other way around, and the reactions of the parts are a result of the overall reaction." Game theory's emphasis on individual choices at the expense of systemic governance matched the neoconservative approach of Reaganism to social problems generally.[23]

Neoliberalism's ideological emphasis on individualism corresponds to a vision of the self as an entity with definite boundaries. Such boundaries, moreover, the state is obligated to respect and defend. On that basis, Reaganism promised to return to individuals "control of their own lives." Whereas ecological thinking emphasized the permeability of boundaries between individuals and systems, a recurring trope of Reaganism was the effort to construct impenetrable defenses or barriers separating internal and external environments. As political theorist Michael Rogin wrote in *"Ronald Reagan," the Movie*, published in 1987, "Having raised anxiety about the permeability of American boundaries, President Reagan split the good within . . . from the bad without."[24]

The prestige of protective barriers forms a characteristic theme of American culture in the eighties, appearing in such diverse manifestations as the Sony Walkman (which allowed you to seal yourself off in an aural cocoon from the sounds of the world around you); and the gated community (which based its promise of security on the faulty premise that all evildoers were outsiders). Reagan's Strategic Defense Initiative likewise offered a radical new strategy for deterring a Soviet nuclear attack. In March 1983 on national television, Reagan called for a program to shoot down Soviet missiles with laser beams to create "an invulnerable shield." Instead of the mutual vulnerability that characterized the previous deterrence strategy of mutually assured destruction, the Strategic Defense Initiative did not attempt to ensure peace, but instead to guarantee invulnerability.[25]

The 1984 film *The River* encapsulates how in the eighties a reemphasis of individualism and defensive barriers captured the popular imagination. Conservative actor Mel Gibson lobbied hard for the starring role, in which he plays Tom Garvey, a struggling family farmer. As William Adams observed at the time, only 3.3 percent of Americans still lived on farms; but the story of the family farm is historically "one of the principal means by which we have imagined and examined our freedom."[26]

The film's agricultural setting serves as a symbol by which it associates virtue with independence. In that sense, the film is a Reagan-era morality play. It both displays and attempts to resolve aesthetically the cultural anxiety created by a new awareness of the realities of interdependence, which had been made clearer to Americans over the past two decades. *The River* engages the problematic relationship of freedom to interdependency and emerges with a regressive solution. As Adams wrote, it resorts to an "older and safer set of understandings" about both freedom and identity, "where the boundaries . . . are sharp and certain" and "where the meanings of independence and dependence . . . are unequivocal." Independence, the movie showed its viewers, must be defended at all costs; because the only alternative is dependency, which means being vulnerable to exploitation by "vast impersonal forces."[27]

The threat of dependency against which Tom Garvey must struggle has both natural and social aspects. The movie begins by showing Tom's corn crop being washed away by a flooding river, and then moves to conflate this disaster with the mysteries of finance capitalism. These are personified in the character of Joe Wade, a farmer-turned-businessman who also carries a torch for Tom's wife, Mae. Significantly, the ecological unsoundness of Tom's continuing effort to farm in a flood zone is introduced in the movie only in order to be explained away by the traditional nature of his identity: "This is my home place," he tells Joe. "My people are buried here." The film places the fault for his insolvency instead on the corrupt society by which he has been surrounded. The corporations, the banks, and the local politicians are all, in a scheme orchestrated by Joe Wade, colluding to rob Tom of his identity by taking away his land. Far from wanting to help him fight off the river's incursions, they want his valley to flood. In fact, they want to flood his valley even more—by building a dam, which will irrigate Joe's 11,000 dry acres but leave Tom's homestead permanently underwater. The film persistently draws together the natural and social forces to which Tom's manly independence is opposed, until the viewer conflates them, as Tom does himself. Tom tells Mae, "I dreamt about the river last night—the river was a big snake, and I couldn't find the head or the tail of it, and it had me wrapped up so tight I couldn't breathe."

The costs of Tom's losing his struggle for independence and falling into dependency are made clear in the film by an interlude in which he temporarily gets a job in a steel mill. Only when he arrives does he realize that he is a virtual captive there. His work is the epitome of alienated labor: the mill is so loud that the men can barely speak to each other, and they don't even know what they are making. In a scene that feels like the emotional center of the movie, a lone deer wanders into this inferno. Curious at first, the men start chasing it, but what begins as an exciting distraction somehow turns dead earnest, as a hundred men chase one frightened doe until they close in on it, surrounding it. The camera shows the viewer close-up shots of the deer's face and of some of the men's. With nowhere to turn, the deer stands immobilized, then it pees forlornly on the floor. Abashed by this reminder of its nature as a sentient biological being, the men back away and open a path for it to flee back to freedom. They recognize something of themselves in her.

The aesthetic sleight-of-hand by which the early scenes conflate the social and natural forces opposing Tom is brought to bear again in the film's climax. This time its purpose is to focus Tom's battle entirely on the river, as he returns to the farm at harvest time to fight again for his independence. Tom's interrelated struggles against Wade Industries, the farm credit crisis, and the erosion of his masculinity are all subsumed into a Manichaean battle he must make against the river that is once again rising. Shoring up the protective barrier between his crops and the river can in this way be presented aesthetically as the answer to every threat.

This allows the classic formulas of Man-against-Nature and Man-against-Man to be substituted for real solutions. Embracing this agenda, Tom's neighbors rally to his aid—the film's only concession to the possibility of a kind of interdependence that does not lead to ruthless exploitation. At one point in the climactic face-off, the film even has Tom defending his homestead against Joe Wade (who has come to try to tear down the levy) with a shotgun, like a character in an old-style Western. "Get off my land!" Tom warns him. In the end, Tom triumphs—not because any of the ecological, economic, or political problems that the film draws on for its emotional power have been solved, but because he bests his rival and piles up enough obstacles to keep an unruly nature and a terrifying dependency temporarily at bay. Thus the movie, like Reaganism generally, clung to a mythical image of the independent man even while acknowledging that it was untenable.

Conclusion: A Metahistory

You got to give off some illumination
or the rest of us be thinkin
that the humans want it all for themselves.
—LEWIS MACADAMS, "CALLIN COYOTE HOME"

What do we gain from a recovery of the history of the culture of feedback? Is there value in knowing that there existed a coherent seventies counterculture that defended with convincing intellectual rigor (despite some ridiculousness) the idea that there were many forms of mind in nature, and in consequence developed a wide range of creative practices for engaging with them? I believe it empowers a necessary change in our assessment of the seventies decade generally.

Culture, as Clifford Geertz famously pronounced in 1973, is constituted by the stories that people tell themselves about themselves.[1] But the stories of the seventies that we have until now had available to us have been warped and fatalistic ones, emphasizing the themes of failure, self-indulgence, and fragmentation. This is because too many historians have taken the satires and self-recriminations of a few seventies pundits too much at face value. Cumulatively, they have built up a myth about the decade's meaning that is becoming as hard to displace as it is inaccurate.

In such histories, the seventies are characterized as first and foremost a decade of malaise. This term originally became widely associated with the decade in 1979, when

President Jimmy Carter's political opponents labeled his televised July 15 address to the nation "the malaise speech."[2] Significantly, however, Carter's speech does not itself contain that word. Correspondingly, the intention of Carter's speech was not to pronounce moral judgment on the decade, but to denounce a lack of moral courage on the part of congressional legislators. Specifically, Carter was urging them to risk offending their lobbyists and constituents and confront the urgent challenge to minimize the country's reliance on outside energy sources, especially Middle Eastern oil. His agenda was therefore very much in line with the ecosystem culture ideal of Gary Snyder and other ecological thinkers of the era, who believed that minimizing the need for imported energy and fossil fuels was a key index of long-term sustainability.

Carter also took the side of ecological thinkers by defining freedom as the ability to navigate with creativity a system-wide change demanded by environmental constraints. The nation's freedom was put to the test, he maintained, by whether Americans could "get together as a nation to resolve our serious energy problem." "True freedom," he proclaimed, was born of "common purpose"; therefore Americans should support more government intervention, both in terms of regulation (of gasoline consumption and fuel imports) and stimulation (of public transportation and alternative energy sources like solar power). If congressional legislators balked at such a proposal, he accused, what stood in the way was "a mistaken idea of freedom" propounded by contemporary free-market advocates: one in which freedom was defined as "the right to grasp for ourselves some advantage over others." To fall prey to this error, Carter warned, would connote a real "crisis of the American spirit."[3]

Unfortunately for Carter and his energy policy agenda, the speech that he hoped would be a rousing call to action was remembered chiefly for its portrayal of this looming failure of will. "In a nation that was proud of hard work, strong families, close-knit communities, and our faith in God, too many of us now tend to worship self-indulgence and consumption," he warned. Although this American jeremiad was centuries old, in his address Carter pinned it to causes that were specific to the past decade. Americans collectively were traumatized, he said; and this was because the years since 1963 had been "years that were filled with shocks and tragedy," including assassinations and attempted assassinations, politicized street violence, the oil crisis, Watergate, Vietnam, and stagflation. The effect of these national traumas ("These wounds are still very deep. They have never been healed," he insisted) was to "fragment" our national identity.[4] In its place, there remained only a narrow self-interest and "a growing disrespect for government."

The oddly psychoanalytic tone of Carter's critique, which attributes a traumatized and dissociative psyche to a personified American nation, can be traced to the influence of Christopher Lasch's 1978 best seller, *The Culture of Narcissism*. Carter's adviser Patrick Caddell, deeply impressed by Lasch's book, had written Carter a long memo in which he summarized Lasch's arguments and urged Carter to use them in his address. As a result, Lasch was invited to a domestic summit organized at Camp David by Carter's communications adviser, and Carter relied heavily on Lasch's ideas in his speech.[5]

Lasch's method in writing *The Culture of Narcissism* harked back to an intellectual form popular in the 1940s, in which psychoanalytic theory was integrated with anthropology or sociology to identify a "national character type." Classics in the field from that earlier era include Ruth Benedict's *The Chrysanthemum and the Sword* and Erik Erikson's *Childhood and Society*. The basic premises of that method had been debunked in the sixties.[6] But Lasch boldly proceeded to identify the national character of the contemporary American as that of the pathological narcissist. The typical American, Lasch wrote, harbored an "antisocial" need for the "immediate gratification" of "cravings" that had "no limits" because they did not come from appetites that could be satisfied, but were driven by a constitutive inability to "find a meaning in life."[7]

Having just finished writing a book lamenting the wane of parental authority in the modern age,[8] and in search of a new but related topic, in the summer of 1976 Lasch had been impressed by Tom Wolfe's intensely subjective critique of seventies popular culture that appeared in *New York Magazine*. Wolfe's essay offered a satirical look at no-fault divorce and marriage counseling, "sexual liberation," "women's liberation," and the Human Potential movement, and cynically summed them all up as a culture of people too fascinated with themselves. *"Let's talk about Me,"* Wolfe asserted, was the hidden transcript of a nationwide and even at times self-righteously "spiritual" quest to find the "Real Me."[9]

Wolfe, who must be seen as a pop artist in the style of Roy Lichtenstein and Andy Warhol, enjoyed a humorous love-hate relationship with the American popular culture that he unrelentingly lampooned. He reveled scatologically in what he saw as the anarchistic diversity of the "Me" Decade, seeing it as the rather anticlimactic culmination of consumer capitalism's historical promise to liberate the average man. "Ordinary folks now had enough money to take it and run off and alter the circumstances of their lives," he wrote. The result was an orgy of self-fulfillment in a mad proliferation of styles, spanning from the radical chic of college kids in combat boots to the freestyle wagon trains of retirees in recreational vehicles. If he found the spectacle heartwarming, he also found it ridiculous. Echoing a critique made earlier by sociologist Philip Rieff in his 1966 book *The Triumph of the Therapeutic*, Wolfe's overall assessment of American popular culture was that it was self-indulgent and lacking in social conscience.

The wide net of satire that Wolfe cast caught some aspects of the culture of feedback in its broad circumference. Hippie communes, Stephen Gaskin, the veneration of "feelings," Zen Buddhism, meditation, and "the ESP movement" all came in for their share of abuse. But Wolfe missed the central, defining role that ecological thinking played in the popularity of these practices, instead interpreting their common denominator as a bout of mystical religiosity that he dubbed the "Third Great Awakening." As a result, he judged them to be socially irresponsible and, ironically, turned the ecological ethic against them himself: accusing the spiritual seekers of the "Me" Decade of a shortsighted self-interest born of consumerism, which made them unable to "conceive of themselves, however unconsciously, as part of a great biological stream." This, Wolfe concluded, led to social fragmentation through a "narcissism" of small differences, as "the faithful split off from one another to seek ever more perfect and refined crucibles in which to fan the Divine spark" of the Real Me.

Lasch's indebtedness to Wolfe's article for the basic fodder of his 1978 book is made clear in an essay that he published in the *New York Review of Books* at the end of September 1976, titled "The Narcissist Society."[10] In this early exploration of his arguments, Lasch cited both Wolfe and Rieff. But he parted company with Wolfe on the question of what the "Me" Decade represented. Wolfe had mocked the somber suggestion that Americans' spiritual searching was an act of despair rather than, as he saw it, the culmination of the corporate-liberal idea of progress. "Newspaper columnists and newsmagazine writers," Wolfe complained,

> theorized that the war in Vietnam, Watergate, the FBI and CIA scandals, had left the electorate shell-shocked and disillusioned and that in their despair the citizens were groping no longer for specific remedies but for sheer faith, something, anything (even holy rolling), to believe in. This was in keeping with the current fashion of interpreting all new political phenomena in terms of recent disasters, frustration, protest, and decline of civilization . . . the Grim Slide.

That hypothesis, Wolfe pointed out, was not borne out by the evidence from polls.[11]

Lasch, however, embraced the etiology that Wolfe had deplored, and argued that narcissism was Americans' nihilistic response to a national trauma caused by Vietnam, stagflation, environmentalism, the decline of the family, and the radical sixties.[12] He claimed to see through the façade of personal growth and expanded awareness to detect the underlying feelings of "loss," "retreat," "despair," and anomie. He wrote:

> It is no secret that Americans have lost faith in politics. The retreat to purely personal satisfactions—such as they are—is one of the main themes of the Seventies. A growing despair of changing society—even of understanding it—has generated on the one hand a revival of old-time religion, on the other a cult of expanded consciousness, health, and personal "growth."
>
> Having no hope of improving their lives in any of the ways that matter, people have convinced themselves that what matters is psychic self-improvement: getting in touch with their feelings, eating health food, taking lessons in ballet or belly dancing, immersing themselves in the wisdom of the East, jogging, learning how to "relate," overcoming the "fear of pleasure." Harmless in themselves, these pursuits, elevated to a program and wrapped in the rhetoric of "authenticity" and "awareness," signify a retreat from the political turmoil of the recent past. Indeed Americans seem to wish to forget not only the Sixties, the riots, the New Left, the disruptions on college campuses, Vietnam, Watergate, and the Nixon presidency, but their entire collective past. . . . To live for the moment is the prevailing passion—to live for yourself, not for your predecessors or posterity. We are fast losing the sense of historical continuity, the sense of belonging to a succession of generations originating in the past and stretching into the future.[13]

Lasch took Wolfe's observations of American popular culture, even his idea of a lost sense of connection to the "great biological stream," and repackaged them with psychoanalytical theory. Then he gave the whole bundle a negative spin. Henry Allen, who interviewed Lasch for the *Washington Post* about the success of his best-selling book, perceived his indebtedness to Wolfe (which by then Lasch was reluctant to acknowledge) and accordingly titled the interview "Doomsayer of the Me Decade." Allen observed that Lasch had taken Wolfe's critique and given it an intellectual pedigree by marrying it to a venerable sociological tradition: "a crowd of other titles advertising decline, collapse [and] discontent" in American culture, dating back to the late 1950s.[14]

Given this, it is hard to see Lasch's book as the unique insight into 1970s American culture that it makes itself out to be. What characterizes the seventies as a cultural period is arguably not the "culture of narcissism" at all. Unfortunately, however, we have until now had only variations on Lasch's diagnosis of national "narcissism" and his etiology of national "trauma" as the stories to tell ourselves about the decade. This template has passed unexamined into all the most-read histories of the decade.

Lasch, in his book, began with the narcissist as a national character type and then traced the "causes" back to a national trauma. Historians typically tell this story in the opposite order, first establishing the idea of a national trauma, and then discerning its "effects" in the various forms of self-seeking that they identify as the heart of seventies culture. The titles of their histories invoke crisis, funk, and nightmare. Making the assumption that key political events were experienced as personal traumas by a great majority of Americans, they characterize the decade as an era of disorganization, confusion, decline, and defeat. Thomas Hine in *The Great Funk*, for example, describes the seventies as a decade when "cars were running out of gas. The country was running out of promise. A President was run out of office. And American troops were running out of Vietnam." Andreas Killen, in *1973 Nervous Breakdown: Watergate, Warhol, and the Birth of Post-Sixties America*, accepts as the premise of his study that the seventies can be equated with "weakness, confusion, and malaise." Killen is among those historians who follow Lasch's lead to the extent of psychopathologizing a personified American nation: Vietnam, Watergate, and the OPEC oil embargo, he writes, "shook the national psyche to its very core." Philip Jenkins, in *Decade of Nightmares*, is another such writer; he characterizes the nation from 1975 onward as being in the grip of a collective "nightmare" provoked by a "sense of pervasive national malaise, decadence, and social failure." The dreamscape that he evokes matches the mise-en-scène of Martin Scorsese's 1976 movie *Taxi Driver*: an image of American cities plagued by violent crime, child pornography, underage prostitutes, and drug deals.[15]

When this bleak stage is established as its historical context, it is easy to imagine the defining culture of the seventies as one in which Americans retreated from social goals, looked out for themselves, and did what they pleased. Tom Wolfe, Killen writes, "put his finger on an important aspect of post-Sixties culture: its promise of a new, radical kind of individualism." Following uncritically in Lasch's footsteps, Bruce Schulman in *The Seventies* likewise maintains that Americans embraced individualism at the expense of "the ideal of social solidarity, the conception of a national community with duties and obligations to one's fellow citizens." Ignoring the profound impact of environmentalism throughout the decade, Schulman argues that "the 1970s witnessed . . . skepticism about large-scale public efforts to remake the world. Economic malaise and political crisis sent the welfare state into retreat and prompted new respect for capitalism." Schulman is a chronicler of the heterosexual southern white male sensibility (his previous books were *Lyndon B. Johnson and American Liberalism* and *From Cotton Belt to Sunbelt*) and his rhetoric reveals a distaste for the image of the seventies that he conjures. He identifies it with "a new ethic of personal liberation" that "trumped older notions of decency, civility, and restraint. Americans widely embraced this looser code of conduct."[16] The popularity of the ecological idea that formed the center of Carter's 1979 speech—namely, that freedom lay specifically in acknowledging and responding creatively to systemic constraints—is not part of Schulman's paradigm.

The grass looks greener to Hine, who is a champion of gay liberation; but both historians are looking at the same distorted field, in which individualism sprouts in the aftermath of devastation. A decade of "disasters," Hine contends, left people "free to invent or reinvent themselves" and to "indulge in" the "pleasure" of "life, in all its messy luxuriance." One could "cease pretending to believe in things that had already failed . . . and live your life in an honest way." But as a result, seventies Americans suffered from "self-absorption and the inability to follow through on commitments." Hine un-ironically echoes the lyrics of a Janis Joplin hit from 1971—"Freedom's just another word for nothin' left to lose"—as a summation of the decade's redefinition of that keyword. The fact that ecological thinking was propagating a different idea of freedom—one in which the individual was a "holon" for whom self-expression was an indispensable ancillary to fulfilling one's social responsibilities—is never even considered.[17]

The logic of the "malaise" story requires that its storytellers imagine a national culture that had "fragmented" (due to trauma) into myriad subcultures by which Americans quested to restore meaning to their lives. Thus Hine describes the seventies as a decade in which the "American Way of Life . . . shattered into a bewildering array of lifestyles," like light reflecting off a disco ball. Schulman, too, discerns a "process of fragmentation," although he less approvingly likens it to a centrifuge, spinning people off willy-nilly in all directions. Jenkins characterizes the culture of the first half of the decade as one in which a high value was placed on diversity. But he does not see this diversity in the way that ecological thinking understood it: as a strategy necessary to the survival of the social system as a whole. Instead, he interprets it through the trope of fragmentation, as the erosion of social coherence into a multitude of distinct ethnic identities.[18]

An important consequence of this "fragmentation" narrative is to rule out the possibility that a coherent alternative to the dominant social order existed or could exist. Whereas ecological thinking envisioned a radically different social order that valued diversity as part of a revised definition of freedom, in the work of these historians, the corporate-liberal social order is cast as the indispensable though invisible solvent that enabled all of America's heterogeneous subcultures to coexist. There are no real alternatives; only alternative lifestyles.

This message is sustained because the "fragmentation" narrative effectually equalizes all cultural forms, whatever their political contents or social meanings, as variations on the same centrifugal search for self-fulfillment. Tom Wolfe saw Jimmy Carter the evangelical Baptist and Jerry Brown the "Zen Jesuit" as cut from the same cloth. Schulman echoes this, seeing the search for meaning taking place on rural communes as a variation on the same impulse that was "filling the churches and hot tubs of the South and Southwest." He identifies it as part of "a newfound respect for religion—a *broad, nationwide interest* in the experience of spiritual rebirth," spanning "from the Baptist revivals of Virginia to the New Age retreats of California" in which "self-realization remained the seeker's objective." Thinking along the same lines originally set out by Wolfe and Lasch, Hine finds little to distinguish among the cultural agendas of the anti-busing Irish, the Symbionese Liberation Army, hippies, feminists, evangelical Christians, gays, environmentalists, and neoconservatives; "all sorts of people came together," he writes, "to try many different approaches to life."[19]

This perspective supports a tendency to misconstrue the ecological thinking of the seventies and minimize its radical potential. Although Hine is dimly aware of the importance of ecological thinking to Americans of the period, he gets its significance only half right, grasping the value it placed on diversity but not the related imperative to organization:

> The science of ecology demonstrates that a diversity of life-forms is a sign of health in an environment. Monocultures are prone to sudden collapse in biological systems, and it's not difficult to extend this insight to social, political, and economic systems as well. Postwar America seemed to aspire to monoculture, and in the seventies it seemed in danger of collapse. The country was clearly going in the new direction: it was time to *break up and try a lot of new paths.*

Reducing ecological thinking to just another search for what Wolfe called the "Real Me" leads these historians to group and classify it with several unrelated forms and practices, making it disappear into the cloud labeled "New Age." Schulman identifies many aspects of the culture of feedback—including rural communes and the *Whole Earth Catalog*; organic food and the ecological ethics of Gary Snyder; and popular interest in the traditional Native American viewpoint—and characterizes them all as aspects of New Age spirituality. He then sides with critics who condemn New Age practices generally as "solipsistic" and a "selfish shirking of social responsibility." Hine likewise groups sustainable living experiments with New Age self-help books like Richard Bach's *Jonathan Livingston Seagull,* and even with white-ethnic identity politics, all of which are encompassed under the general rubric of "consciousness"; as he observes, "there were many kinds of consciousness emerging."[20]

This distorted version of the seventies has by extension distorted our historical understanding of the sixties. Jenkins asserts that "the story of American culture in the early 1970s can be seen as the mainstreaming of sixties values, the point at which countercultural ideas reached a mass audience."[21] This may be true; but what "sixties values" and "countercultural ideas" are being referred to?

Jenkins sees the sixties counterculture as characterized primarily by "permissiveness." He argues that its permissive attitudes toward sexuality, drug use, clothing, hairstyles, and religiosity percolated through to the mainstream culture in the seventies. Schulman also identifies the "countercultural ethos" with a permissiveness that is implicitly polluting, as "curse words ceased to shock . . . [and] states relaxed or repealed obscenity laws, abortion restrictions, and regulations prohibiting the sale of contraceptives." He sums up the cultural legacy of the sixties as "music, nakedness, and drugs"; it is not hard to see in this a paraphrase of Ian Dury and the Blockheads' 1977 hit single, "Sex and Drugs and Rock and Roll." In these histories, the cultural legacy of the counterculture and its negative effects are typically encapsulated by a mini-history in which the utopianism of Woodstock blazes the way to the ugliness of Altamont. Lasch wrote in *The Culture of Narcissism* that "the zonked-out lovefest of the 'Woodstock Nation' deteriorated into the murderous chaos of Altamont." Schulman echoes this in writing that "Altamont swept into the open all the ugly features of the counterculture."[22]

The counterculture thus becomes responsible for tempting Americans down the path of narcissism in a moment when they are numbed by trauma. Attributed to the counterculture is an attitude of escapism that leads eventually to social isolation. In a subchapter titled "The Legacy of Woodstock," Schulman quotes Country Joe's disparaging assessment of his fans: "You take drugs, you dance around, you build yourself a fantasy world where everything's beautiful." In *The Great Funk*, Hine writes that an emphasis on personal growth ultimately pushed people into living alone. "One reason that people began talking to their plants may be that there was nobody else at home with whom you could have a conversation," he muses.[23]

Because the counterculture is identified only with permissiveness and escapism, the radicalism of the sixties is shown to be a dead end. This lesson is imparted by summarizing the history of the New Left as the myth of Icarus: the story of an unrealistic youthful idealism that plummets into despair. Ignoring the comparatively quiet radicalism of the ecological revolution that continued into the seventies and beyond, the extant histories focus on sensational news events featuring violent confrontations to tell the story of sixties radicalism and its fall. Thus Killen maintains that "the final spasm of the Sixties came with the guerrilla organizations of the early 1970s, exemplified by the SLA."[24]

To associate radicalism with an excessive idealism of youth implies that it is rooted in immaturity. Schulman writes of the "disrespect" and "contempt for authority" exhibited by "young radicals" of "a self-described New Left." He uncritically quotes the president of Columbia University (whose office was at the time being occupied by activists) in asserting that "they have taken refuge in a turbulent and inchoate nihilism." This then becomes a lens through which to view any aspects of radicalism in the seventies decade, including the call for a post-Cartesian epistemology. "Not just the government, but *all* sources of authority became targets for distrust and mockery," Schulman writes. "Even science, the triumphant force that had landed man on the moon."[25]

The main point that these histories make about the radical possibilities of the sixties, however, is that by the early seventies they had ended. This is the only conclusion consistent with the master narrative of failure and self-seeking that they have adopted from Wolfe and Lasch. In his 1976 essay, Wolfe quoted Barbara Garson, a disillusioned and "bitter" former member of the New Left, who told him that at the end of the sixties the leaders of the New Left "forsook everything" political and disappeared to pursue personal agendas of spirituality and self-help. In his essay "The Narcissistic Society," Lasch followed suit and described the former radicals as now shopping in "spiritual supermarkets." The idea is echoed uncritically by seventies historians. Killen, for example, contends that by 1973 "radicalism had run its course. One by one the leaders of the counterculture fell off the radar screen." Schulman similarly asserts that "the New Left, *the radical movement envisioning real change,* fizzled after Chicago [the Democratic Convention of 1968]." "Losing focus," he writes, it "descended into decadence" and "fell apart" as it "embraced new concerns"—among them, ethnic diversity, feminism, intersubjective dialogue, body-mind holism, and ecology. He cannot see the radicalism in these. He concludes, "The counterculture expanded in the Seventies, spreading a less formal, more open and freewheeling way of life. But the real efforts at cultural revolution, at creating a sustainable alternative, collapsed or became diluted."[26]

Characterizing radicalism as a kind of immaturity facilitates this story because it enables a narrative in which radicalism is seen to disappear in the course of a natural progression toward maturity. Its demise does not require explanation in terms of the dynamics of its political defeat. It can be told instead as a generational coming-of-age story. Thus Hine portrays the radicalism of the sixties as giving way to tolerance in the seventies. "The sixties were about struggle," he writes; "the seventies about acceptance." This was because the newly adult baby boomers in their "childbearing years" traded in their fiery idealism to become interested in personal growth; hence domestic interiors emphasizing "the warm [and] the fuzzy" and "jungles of spider plants and ferns to serve as refuges from the world outside." Jenkins takes a similar line in *Decade of Nightmares,* explaining that while radicalism marked the baby boomers' reckless teenage years, by the late seventies they were having babies of their own and wanted to "reassert parental authority" and "restrain . . . hedonistic impulses," thus laying the foundations for Reaganism. "When the baby boomers were in their late teens in 1968, many wanted to expand their *own* rights and rejected official restrictions on sexual activity and drug use," Jenkins writes. "A decade later, the same individuals were starting families and were *naturally* concerned about the welfare of their children."[27] The implied subtext is that both cultural movements—sixties "permissiveness" and eighties neoconservatism—were equally manifestations of a self-serving "Me" generation.

As Jenkins's argument illustrates, the final disservice done by the existing histories of the seventies is to misrepresent the meaning of Reaganism's rise to dominance. Once it has been established that the seventies are the decade of "Me" and "malaise," it is logical to imagine Reaganism either as an inevitable repudiation of the moral lassitude of the previous decade, or as the natural culmination of its individualistic ethos. Sometimes, somewhat contradictorily, it is argued to be both. These explanations omit the possibility that in the early 1980s a viable cultural alternative represented by the culture of feedback lost its battle against the better-organized and better-funded forces of conservatism.

Many histories explain Reaganism as a natural swing of the cultural pendulum. Jenkins writes that "alternating cycles of hedonism and puritanism have occurred throughout American history, commonly focusing on attitudes towards sexuality and drink or substance abuse." He finds it convenient again in this context to personify the nation, declaring, "Binge decades are followed by collective hangovers, eras of 'clean and sober.'" Killen, pursuing his psychoanalytic trope, argues that repeatedly in history the restoration of a monarch (Reagan, in this case) follows hard upon the heels of a patricidal revolution (the sixties), since "the intense need to cast off one form of authority was invariably accompanied by an equally intense search for new forms of it." Hine suggests that "freedom . . . was simply exhausting. That's why, in 1981, Americans reached back in time to old, simple messages and pretended to believe them."[28]

Although it would seem self-contradictory, these same historians also argue that nothing changed with the advent of Reaganism, because the culture of the seventies harbored no other values than those that Reaganism espoused. Thus Hine reluctantly embraces the Sony Walkman as the culmination of what the seventies had been aiming for all along: "Each person had his own music," he explains, "at least others weren't being bothered." Jenkins similarly argues that there was no such thing as a Reagan revolution in the early 1980s. "We should see these political changes less as a retreat or reaction than as a natural culmination of earlier liberalism . . . a follow-through," he insists. Armed with this premise, he finds the seeds of eighties conservatism everywhere in the anti-establishment movements of the seventies. Feminists' anti-rape activism prepared the ground for conservative politicians to plant baseless fears of sexual predators. "A tradition of conspiracy theory handed down by Watergate-era liberals" paved the way for eighties hysteria about Satanism in popular song lyrics. The "dualistic ideas of good and evil and visions of imminent disaster" that characterized the millennialism of President Reagan were carried over from "the doomsday environmentalist prophecies of the early 1970s." The list is potentially endless, because it is history written by constructing analogies rather than by tracing genealogies.[29]

Schulman also argues that "the early 1980s did not so much repudiate the political and cultural legacies of the Seventies as complete and consolidate them." Because of his undue focus on the culture of white southerners, he sees "a conservative, Southern ascendancy" as the leading event of the seventies decade—a bias that leads him to cast Reaganism's triumph in 1984 as the result of a gradual but inevitable progression. As a result, his narrative frames the eclipse of ecological thinking as an emerging national consensus rather than as the outcome of a political power struggle. "Increasingly, *all sorts of Americans, even those with dreams of radical reform, looked to the entrepreneur and the marketplace* as the agent of national progress and dynamic social change," he recounts. But his only evidence for this claim are statements by Nixon and Reagan.[30]

Histories that see Reaganism and its free-market ethos as already implicit in the countercultural attitudes of the seventies cede to Reaganism too much cultural terrain in view of what must, in the long run, prove only a short-term victory. Although Reaganism did prevail in the mid-1980s, discrediting much of the culture of feedback and bending mainstream attitudes away from environmentalism and ecological thinking, its triumph was not an inevitable fate. To the contrary, in the context of the longer trajectory of American modernity and postmodernity, it can be seen to constitute but one beat in an ongoing debate about the best future course of the nation, organized around cultural keywords like "freedom," "democracy," and "efficiency." When we look into the history of American culture since 1984, newly equipped with a knowledge of how to spot the continued influence of the culture of feedback, the partial nature of Reaganism's victory is readily revealed.[31]

In this broader context, the faulty story of the seventies as the "malaise" that brought us to Reaganism has problematic implications for the historiography of postmodernism in general. The question of how to define the postmodern condition affects how we think of our present moment and of what it means to be living in it. Until now, two particular understandings of postmodernity—those rooted in the ideas of Jean Baudrillard and Michel Foucault, respectively—have predominated as the favorite master narratives of contemporary historians. Both of these partake of what Eve Kosofsky Sedgwick in her book *Touching Feeling: Affect, Pedagogy, Performativity* calls a "paranoid reading." The Freudian descriptor is used here not in the crude sense of defining a national psyche, but to characterize a persistent pattern of practice shared by many cultural critics. Because "paranoia requires that bad news be always already known," paranoid critiques are those that frame every cultural development as another turn for the worse, or as a Trojan horse promising good things but bearing within it the forces that will further erode our identity and freedom. The cumulative effect, Sedgwick writes, is to foster an "everyday, rather incoherent cynicism."[32]

The definition of postmodernity proposed by Baudrillard positions us as trapped in a hall of mirrors produced by the proliferation of communications media. The only recourse is a self-ironizing "parody [that] renders submission and transgression equivalent."[33] This is the version of postmodernity that Killen adopts in framing his book *1973 Nervous Breakdown*. For him the quintessential seventies American self is therefore the self-consciously media-saturated, "performative" and "schizophrenic" self of Andy Warhol, Patty Hearst, and the stars of *An American Family*. The ultimate futility of its countercultural aspirations is embodied in the "revolutionary kitsch" of Patty Hearst's turnstile of identities (heiress, freedom fighter, and media celebrity) and Andy Warhol's silk-screen portraits of Chairman Mao. From these Killen derives his paranoid lesson:

> This transformation of a hero of the New Left into an ambiguous kind of celebrity image offers a comment on the fate of the Sixties culture of radicalism. Even in the throes of crisis capitalism was still capable of swallowing all revolutionary impulses . . . [and accomplishing the] commodification of dissent.[34]

This is a postmodernism stuck in the (revolving) doorway of modernity, capable of ironizing the effects of the modern social order but incapable of formulating a viable alternative.

Although specifically indebted to the ideas of Baudrillard, this story may be seen as one variation on a pervasive critical trope that is most often associated with the writings of Foucault. In such histories, cultural developments that were initially hailed as liberating are revealed to be just more subtle forms of entrapment. A version of this story forms the basis for Lasch's *Culture of Narcissism* (since psychopathology can be thought of as a kind of trap) and thus for all the histories of the seventies that Lasch's work informs. This kind of paranoid reading also underpins the arguments of a batch of scholarly works that see systems theory (cast as "cybernetics") as leading only to more intense forms of commodification, greed, and dehumanization. Among these are Philip Mirowski's *Machine Dreams*, Orit Halpern's *Beautiful Data*, and Paul Edwards's *The Closed World*. These all see cybernetics as leading inevitably to the more totalizing exercise of what Foucault called "biopower": a world of intensive surveillance, dwindling options, and compulsory interfaces with cyborg technology.[35]

As Sedgwick writes:

> Paranoia knows some things well and others poorly. . . . In a world where no one need be delusional to find evidence of systemic oppression, to theorize out of anything but a paranoid critical stance has come to seem naïve, pious, and complaisant. . . . [B]ut it seems to me a great loss when paranoid inquiry comes to seem entirely coextensive with critical theoretical inquiry rather than being viewed as one kind of cognitive/affective theoretical practice among other, alternative kinds.

Outside the paranoid fold, she continues, "because there can be terrible surprises . . . there can also be good ones. Hope, often a fracturing, even a traumatic thing to experience, is among the energies by which the reparatively positioned reader tries to organize the fragments and part objects she encounters."[36]

As opposed to the available dystopian histories, then, the reader may find it hopeful to discover that there are other postmodernisms besides those imagined by the followers of Foucault and Baudrillard. The culture of feedback offers a post-humanism that emphasizes not cyborgs but the common ground of subjectivity among humans, animals, and plants. Its critique of the notions of objectivity and progress enshrined in modern science does not point toward either relativism or ignorance, but to an epistemology and ethics of intersubjectivity, which issues in the caring attentiveness of empathy and the "performative" engagement of walking a mile in another person's shoes.[37]

When seen in conjunction with the earlier culture of spontaneity, the seventies culture of feedback represents a remarkably continuous popular culture spanning more than half a century, dedicated to the principles of intersubjectivity and body-mind holism. This is the primary cultural legacy bequeathed by the sixties to the seventies. It was not limited to New Age spirituality nor was it wholly, or even primarily, self-centered or consumerist. On the contrary, it was "responsible," as participants in the culture of feedback understood that idea on their own terms consistent with the ecological ethic. For all its ludicrous aspects, it constitutes a "usable past" inviting hope and not despair.[38]

Notes

INTRODUCTION

1. F. H. George, *Cybernetics and the Environment* (London: Elek Books, 1977), 66; V. L. Parsegian, *This Cybernetic World of Men, Machines, and Earth Systems* (New York: Doubleday, 1972), 158–59; Ludwig von Bertalanffy, *General System Theory: Foundations, Development, Applications*, rev. ed. (New York: George Braziller, 1968), 3; Erich Jantsch, *The Self-Organizing Universe: Scientific and Human Implications of the Emerging Paradigm of Evolution* (New York: Pergamon Press, 1980), 19; Robert Leo Smith, *The Ecology of Man: An Ecosystem Approach* (New York: Harper & Row, 1972), 3–4.
2. David Oates, *Earth Rising: Ecological Belief in an Age of Science* (Corvallis: Oregon State University Press, 1989), 3, 5. On the postwar culture of spontaneity, see Daniel Belgrad, *The Culture of Spontaneity: Improvisation and the Arts in Postwar America* (Chicago: University of Chicago Press, 1998).
3. Arne Næss, "The Shallow and the Deep, Long-Range Ecology Movement: A Summary," *Inquiry* 16, no. 1 (February 1973): 95.
4. Paul Shepard, "Ecology and Man: A Viewpoint," in *The Ecological Conscience: Values for Survival*, ed. Robert Disch (Englewood Cliffs, NJ: Prentice-Hall, 1970), 60, 63–64. On popular culture as intellectual history, see Warren Susman, *Culture as History: The Transformation of American Society in the Twentieth Century* (New York: Pantheon, 1984), 272–73. On cultural forms as socially meaningful patterns of cultural practice, see Clifford Geertz, *The Interpretation of Cultures* (New York: Basic Books, 1973), 449–50; and Daniel Belgrad, "The Gnostic Heritage of Heavy Metal," *Culture and*

Religion 17, no. 3 (September 2016): 288. Affect is the physiological "feeling" response of the human organism to experiences: see Julia Kristeva, *The Kristeva Reader*, ed. Toril Moi (New York: Columbia University Press, 1986), 19, 94–97, 103–5, 120–22; Eve Kosofsky Sedgwick, *Touching Feeling: Affect, Pedagogy, Performativity* (Durham, NC: Duke University Press, 2003), 97–117; Brian Massumi, *Parables for the Virtual: Movement, Affect, Sensation* (Durham, NC: Duke University Press, 2002); Nigel Thrift, "Intensities of Feeling: Towards a Spatial Politics of Affect," *Geografiska Annaler* 86B, no. 1 (2004): 57–78; and Ben Highmore, "Bitter After Taste: Affect, Food, and Social Aesthetics," in *The Affect Theory Reader*, ed. M. Gregg and G. Seigworth (Durham, NC: Duke University Press, 2010), 123.

5. David Harvey, *A Brief History of Neoliberalism* (Cambridge: Oxford University Press, 2005), 37–38; Shepard, "Ecology and Man," 59. On political keywords, see Daniel Rodgers, *Contested Truths: Keywords in American Politics since Independence* (New York: Basic Books, 1987), 5–11, 214.
6. Some well-known approaches to describing postmodernism include Frederic Jameson, *Postmodernism, or, The Cultural Logic of Late Capitalism* (Durham, NC: Duke University Press, 1990); and David Harvey, *The Condition of Postmodernity: An Enquiry into the Origins of Cultural Change* (Cambridge, MA: Blackwell, 1989). See also Charles Jencks, ed., *The Post-Modern Reader* (New York: St. Martin's Press, 1992).
7. Belgrad, *Culture of Spontaneity*, 110–13, 123–27.
8. Fred Turner, *From Counterculture to Cyberculture: Stewart Brand, the Whole Earth Network, and the Rise of Digital Utopianism* (Chicago: University of Chicago Press, 2006), 45, 73, 79–80, 90, 92, 97, 99–104.
9. Gary Snyder, *The Real Work: Interviews and Talks, 1964–1979* (New York: New Directions, 1980), 90.
10. Stewart Brand, "Both Sides of the Necessary Paradox," *Harper's*, November 1973, 20.
11. Peter Russell, *The Global Brain: Speculations on the Evolutionary Leap to Planetary Consciousness* (Boston: Houghton Mifflin, 1983), 97, 142, 155, 167–72, 216.
12. Eduardo Kohn, *How Forests Think: Toward an Anthropology Beyond the Human* (Berkeley: University of California Press, 2013), 159. See also Michael Marder, *Plant-Thinking: A Philosophy of Vegetal Life* (New York: Columbia University Press, 2013); Kari Weil, *Thinking Animals: Why Animal Studies Now?* (New York: Columbia University Press, 2012); Donna Haraway, *Staying with the Trouble: Making Kin in the Chthulucene* (Durham, NC: Duke University Press, 2016); Brian Massumi, *What Animals Teach Us about Politics* (Durham, NC: Duke University Press, 2014); Frans de Waal, *Are We Smart Enough to Know How Smart Animals Are?* (New York: Norton, 2017); and Thom van Dooren, "Mourning Crows: Grief and Extinction in a Shared World," in *Routledge Handbook of Human-Animal Studies*, ed. Garry Marvin and Susan McHugh (New York: Routledge, 2014), 275–89.

13. Frederick Winslow Taylor, *The Principles of Scientific Management* (New York: Norton, 1967), 6–12, 28. For an account of culture as the site of political struggle, see T. J. Jackson Lears, "The Concept of Cultural Hegemony," *American Historical Review* 90, no. 3 (June 1985): 571–72. On the meaning of efficiency in Fordism and Taylorism, see Samuel Haber, *Efficiency and Uplift: Scientific Management in the Progressive Era, 1890–1920* (Chicago: University of Chicago Press, 1964).
14. Taylor, *Principles of Scientific Management*, 26, 31, 36–39, 43, 70; Richard Edwards, *Contested Terrain: The Transformation of the Workplace in the Twentieth Century* (New York: Basic Books, 1979), 114, 122.
15. Taylor, *Principles of Scientific Management*, 31. See Peter J. Schmitt, *Back to Nature: The Arcadian Myth in Urban America* (Oxford: Oxford University Press, 1969); and Harvey Green, *Fit for America: Health, Fitness, Sport, and American Society* (New York: Pantheon, 1986). See Richard Slotkin, *Gunfighter Nation: The Myth of the Frontier in Twentieth-Century America* (New York: Atheneum, 1992).
16. See Neil Maher, "'Crazy Quilt Farming on Round Land': The Great Depression, the Soil Conservation Service, and the Politics of Landscape Change on the Great Plains during the New Deal Era," *Western Historical Quarterly* 31, no. 3 (Autumn 2000): 320, 323, 332, 335; Paul B. Sears, *Deserts on the March* (Norman: University of Oklahoma Press, 1935), 144 (emphasis mine); Curt D. Meine, *Aldo Leopold: His Life and Work* (Madison: University of Wisconsin Press, 1988), 313–14; Susan L. Flader, *Thinking Like a Mountain: Aldo Leopold and the Evolution of an Ecological Attitude toward Deer, Wolves, and Forests* (Madison: University of Wisconsin Press, 1974); Susan L. Flader and J. Baird Callicott, "Introduction," in *The River of the Mother of God and Other Essays by Aldo Leopold* (Madison: University of Wisconsin Press, 1991), 4–8.
17. Aldo Leopold, "The Conservation Ethic," in Disch, *Ecological Conscience*, 45; Sears, *Deserts*, 227, 2–3, 141.
18. Sears, *Deserts*, 87, 90, 3.
19. Anthony Wilden, *System and Structure: Essays in Communication and Exchange*, 2nd ed. (New York: Tavistock, 1980), xlii, 463–64.
20. Shepard, "Ecology and Man," 57.
21. Næss, "The Shallow and the Deep," 5.
22. W. Ross Ashby, *Design for a Brain: The Origin of Adaptive Behaviour*, 2nd rev. ed. (New York: John Wiley & Sons, 1960), 40.
23. See Sean Wilentz, *Chants Democratic: New York and the Rise of the American Working Class, 1788–1850* (New York: Oxford University Press, 1984); and John L. Thomas, *Alternative America: Henry George, Edward Bellamy, Henry Demarest Lloyd, and the Adversary Tradition* (Cambridge, MA: Belknap Press, 1983).
24. Paul A. Weiss, *The Science of Life: The Living System—A System for Living* (Mt. Kisco, NY: Futura, 1973), 58. See Daniel Belgrad, "Democracy, Decen-

tralization, and Feedback," in *American Literature and Culture in an Age of Cold War: A Critical Reassessment*, ed. Steven Belletto and Daniel Grausam (Iowa City: University of Iowa Press, 2012), 59–60.

25. George Lipsitz, *Rainbow at Midnight: Labor and Culture in the 1940s* (Urbana: University of Illinois Press, 1994), 57, 61. Dwight Macdonald wrote in 1946 that fascism, communism, and monopoly capitalism were all versions of the same centralized social structure, which he called "bureaucratic collectivism." Dwight Macdonald, "The Root Is Man (Part I)," *Politics* 3 (April 1946): 112. See also Les Adler and Thomas Paterson, "Red Fascism: The Merger of Nazi Germany and Soviet Russia in the American Image of Totalitarianism, 1930's–1950's," *American Historical Review* 74, no. 4 (1970): 1046–64. Margaret Mead, "The Comparative Study of Culture and the Purposive Cultivation of Democratic Values," in *Science, Philosophy, and Religion, Second Symposium*, ed. Lymon Bryson and Louis Finkelstein (New York: Columbia University, 1942); Gregory Bateson, *Steps to an Ecology of Mind: Collected Essays in Anthropology, Psychiatry, Evolution, and Epistemology* (Chicago: University of Chicago Press, 1972), 161–62. See also David Lipset, *Gregory Bateson: The Legacy of a Scientist* (Englewood Cliffs, NJ: Prentice-Hall, 1980), 166.
26. Bateson, *Steps*, 159–60.
27. Bateson, *Steps*, 164–67.
28. Steve J. Heims, *John von Neumann and Norbert Wiener: From Mathematics to the Technologies of Life and Death* (Cambridge, MA: MIT Press, 1980), 302, 184, 202–3, 215–16; Steve Joshua Heims, *The Cybernetics Group* (Cambridge, MA: MIT Press, 1991), 28.
29. Fred Turner, *The Democratic Surround: Multimedia and American Liberalism from World War II to the Psychedelic Sixties* (Chicago: University of Chicago Press, 2013), 257, 161.
30. Robert McIntosh, *The Background of Ecology: Concept and Theory* (Cambridge: Cambridge University Press, 1985), 164, 168, 170, 181; Aldo Leopold, *A Sand County Almanac* (New York: Oxford University Press, 1966), 140–41; Donald Worster, *Nature's Economy: A History of Ecological Ideas*, 2nd ed. (New York: Cambridge University Press, 1994), 270.
31. Leopold, *Sand County Almanac*, 140–41.
32. Lipset, *Gregory Bateson*, 262; Turner, *From Counterculture to Cyberculture*, 38; Heims, *Cybernetics Group*, 30. Among the most influential critics of centralization in New Left circles were C. Wright Mills and Paul Goodman. See C. Wright Mills, "The Structure of Power in American Society," *British Journal of Sociology* 9, no. 1 (March 1958): 31; Mills, *The Power Elite* (New York: Oxford University Press, 1956); and Paul Goodman, *People or Personnel: Decentralizing and the Mixed System* (New York: Random House, 1965).
33. Gary Lachman, *Turn Off Your Mind: The Mystic Sixties and the Dark Side of the Age of Aquarius* (New York: Disinformation Press, 2001), 348–53; Bateson, *Steps*, 440.

34. Belgrad, *Culture of Spontaneity*, 29–31, 35–37, 56–62; Colin Campbell, *The Romantic Ethic and the Spirit of Modern Consumerism* (New York: Blackwell, 1987).
35. Cynthia Novack, *Sharing the Dance: Contact Improvisation and American Culture* (Madison: University of Wisconsin Press, 1990), 188.
36. Brian Eno, *Imaginary Landscapes* (New York: Mystic Fire Video, 1989); Ian McHarg, *Design with Nature* (Garden City, NY: Natural History Press, 1969), 121–25.

CHAPTER ONE

1. N. Katherine Hayles, *How We Became Posthuman: Virtual Bodies in Cybernetics, Literature, and Informatics* (Chicago: University of Chicago Press, 1999), 10.
2. Paul Brooks, "Notes on the Conservation Revolution," in *Ecotactics: The Sierra Club Handbook for Environmental Activists*, ed. John Mitchell (New York: Pocket Books, 1970), 36–37; Frank Benjamin Golley, *A History of the Ecosystem Concept in Ecology: More than the Sum of the Parts* (New Haven, CT: Yale University Press, 1993), 62–66, 104–5.
3. Gregory Bateson, *Steps to an Ecology of Mind: Collected Essays in Anthropology, Psychiatry, Evolution, and Epistemology* (Chicago: University of Chicago Press, 1972), 512, 467; *Galatians* 6:7, English Standard Version.
4. Robert McIntosh, *The Background of Ecology: Concept and Theory* (Cambridge: Cambridge University Press, 1985), 161, 164, 168, 199, 181; Steve Joshua Heims, *The Cybernetics Group* (Cambridge, MA: MIT Press, 1991), 27; Ramón Margalef, *Perspectives in Ecological Theory* (Chicago: University of Chicago Press, 1968), 3; Peter J. Taylor, "Technocratic Optimism, H. T. Odum, and the Partial Transformation of Ecological Metaphor after World War II," *Journal of the History of Biology* 21, no. 2 (1988): 214, 218, 225. See G. Evelyn Hutchinson, "Circular Causal Systems in Ecology," *Annals of the New York Academy of Sciences* 50 (1948): 221–46. Sharon E. Kingsland, *The Evolution of American Ecology, 1890–2000* (Baltimore: Johns Hopkins University Press, 2005), 194.
5. McIntosh, *Background of Ecology*, 199–200, 202; Kingsland, *Evolution of American Ecology*, 185–86; Howard T. Odum and Richard C. Pinkerton, "Time's Speed Regulator: The Optimum Efficiency for Maximum Power Output in Physical and Biological Systems," *American Scientist* 43, no. 2 (April 1955): 331.
6. David Oates, *Earth Rising: Ecological Belief in an Age of Science* (Corvallis: Oregon State University Press, 1989), 116; Odum and Pinkerton, "Time's Speed Regulator," 333.
7. Odum and Pinkerton, "Time's Speed Regulator," 336. See Peter Farb, *Ecology* (New York: Time, Inc., 1963); and Frank Egler, "Pesticides—In Our Ecosystem," *American Scientist* 52, no. 1 (March 1964): 110–36.

8. Odum and Pinkerton, "Time's Speed Regulator," 342; Paul B. Sears, *Deserts on the March* (Norman: University of Oklahoma Press, 1935), 9, 90, 137.
9. Sears, *Deserts*, 9, 20; Odum and Pinkerton, "Time's Speed Regulator," 343.
10. Kingsland, *Evolution of American Ecology*, 194, 210; Margalef, *Perspectives in Ecological Theory*, 29.
11. D. Ramon Margalef, "Information Theory in Ecology," *General Systems: Yearbook for the Society for the Advancement of General Systems Theory* 3 (1958): 46.
12. Margalef, *Perspectives in Ecological Theory*, 18–19.
13. In 1944 John von Neumann published *Theory of Games and Economic Behavior*, coauthored with economist Oskar Morgenstern. Steve J. Heims, *John von Neumann and Norbert Wiener: From Mathematics to the Technologies of Life and Death* (Cambridge, MA: MIT Press, 1980), 291–92; William Poundstone, *Prisoner's Dilemma: John von Neumann, Game Theory, and the Puzzle of the Bomb* (New York: Doubleday, 1992), 6.
14. In addition, experiments conducted throughout the 1950s and 1960s showed that in games played by people rather than computers, interpersonal factors like ego and personality consistently trumped rationality as the basis of decision making, making people much more unpredictable than the equations suggest. Heims, *von Neumann and Wiener*, 81–82, 85, 296; Poundstone, *Prisoner's Dilemma*, 64, 97, 99, 116, 118, 172–73, 176, 260–61.
15. Oliver G. Haywood Jr., *Military Doctrine of Decision and the Von Neumann Theory of Games*, RM-528 (Santa Monica, CA: RAND Corporation, 1951), 1–7; Poundstone, *Prisoner's Dilemma*, 8, 130–32; Sharon Ghamari-Tabrizi, *The Worlds of Herman Kahn: The Intuitive Science of Thermonuclear War* (Cambridge, MA: Harvard University Press, 2005), 52; Heims, *von Neumann and Wiener*, 316.
16. Bateson, quoted in Poundstone, *Prisoner's Dilemma*, 168; Ghamari-Tabrizi, *Worlds of Herman Kahn*, 17; Herman Kahn, *On Thermonuclear War* (New York: Taylor and Francis, 2017 [1960]), xvi, 19–21. See also Herman Kahn, *The Nature and Feasibility of War and Deterrence* (Santa Monica, CA: Rand Corporation, 1960), 39–40, https://rand.org/content/dam/rand/pubs/papers/2005/P1888.pdf.
17. Heims, *von Neumann and Wiener*, 296, 301, 307–8; Norbert Wiener, *Cybernetics: Or Control and Communication in the Animal or Machine*, 2nd ed. (Cambridge, MA: MIT Press, 1961), 159. Wiener wrote that his antagonist was an "Augustinian" devil, whereas von Neumann's was a "Manichaean devil." Norbert Wiener, *The Human Use of Human Beings: Cybernetics and Society*, 2nd rev. ed. (Boston: Houghton Mifflin, 1954), 34–38.
18. Poundstone, *Prisoner's Dilemma*, 44–48; Heims, *von Neumann and Wiener*, 296; Paul A. Weiss, "The System of Nature and the Nature of Systems," in *Toward a Man-Centered Medical Science*, ed. Karl E. Schaefer, Herbert Hensel, and Ronald Brady (Mt. Kisco, NY: Futura, 1977), 19–20 52–53; Heims, *Cybernetics Group*, 246.

19. Heims, *Cybernetics Group*, 98–99, 109–10.
20. Heims, *von Neumann and Wiener*, 89, 297; Wiener, *Cybernetics*, 33, 38.
21. Wiener, *Human Use* (1954), 50–51, 122; Heims, *von Neumann and Wiener*, 301; Wiener, *Cybernetics*, 29, 158–59.
22. Wiener, *Human Use* (1954), 43, 45–46; Norbert Wiener, *The Human Use of Human Beings: Cybernetics and Society*, 1st ed. (Boston: Houghton Mifflin, 1950), 35, 37; this language was omitted from the second edition.
23. Wiener, *Human Use* (1954), 48, 50–52. Arne Næss similarly wrote in his 1973 essay explicating the principles of deep ecology, "Diversity enhances the potentialities of survival, the chances of new modes of life, the richness of forms. . . . Ecologically inspired attitudes therefore favour diversity of human ways of life, of cultures, of occupations, of economies." Arne Næss, "The Shallow and the Deep, Long-Range Ecology Movement: A Summary," *Inquiry* 16, no. 1 (February 1973): 96.
24. Wiener, *Human Use* (1954), 128–29, 181–82; Norbert Wiener, *God and Golem, Inc.: A Comment on Certain Points Where Cybernetics Impinges on Religion* (Cambridge, MA: MIT Press, 1964), 92, 84; Wiener, *Cybernetics*, 171; Gunnar Myrdal, *Against the Stream: Critical Essays on Economics* (New York: Pantheon, 1973), 143, 323.
25. Kahn, *On Thermonuclear War*, xx, 403–6; Tom Wells, *Wild Man: The Life and Times of Daniel Ellsberg* (New York: Palgrave, 2001), 148; Poundstone, *Prisoner's Dilemma*, 204.
26. Daniel Ellsberg, *Papers on the War* (New York: Simon & Schuster, 1972), 42, 44, 48, 71, 102–5, 122, 132–35.
27. Dan Ellsberg, "The First Two Times We Met," in *Gary Snyder: Dimensions of a Life*, ed. Jon Halper (San Francisco: Sierra Club Books, 1991), 334–36, 338.
28. Daniel Belgrad, *The Culture of Spontaneity: Improvisation and the Arts in Postwar America* (Chicago: University of Chicago Press, 1998), 199, 256–57; J. Michael Mahar, "Scenes from the Sidelines," in *Gary Snyder*, ed. Halper, 12–13, 15; Gary Snyder, *The Real Work: Interviews and Talks, 1964–1979* (New York: New Directions, 1980), 95, 162; Gary Snyder, *The Gary Snyder Reader* (Washington, DC: Counterpoint, 1999), 612; Philip Yampolsky, "Kyoto, Zen, Snyder," in *Gary Snyder*, ed. Halper, 66–67; Pat Smith and Mariana Gosnell, "That Snyder Sutra," in *Ecotactics*, ed. Mitchell, 87.
29. Gary Snyder, *The Old Ways: Six Essays* (San Francisco: City Lights Books, 1977), 20–21, 60–61. Snyder adopted his terminology from a talk that environmentalist Raymond Dasmann gave at Cambridge University in December 1974; see Raymond Dasmann, *Called by the Wild: The Autobiography of a Conservationist* (Berkeley: University of California Press, 2002), 152.
30. Snyder, *Real Work*, 130, 28, 69.
31. Snyder, *Real Work*, 89–90, 141.
32. Snyder, *Old Ways*, 93, 20–21, 60–61.

33. McIntosh, *Background of Ecology,* 310. See Margalef, *Perspectives in Ecological Theory,* 45–48. The first annual report of the Council on Environmental Quality in 1970 paraphrased the findings of Margalef and the Odums, emphasizing biodiversity and feedback loops as key factors in nature's resiliency in the face of disruption: "The more interdependencies in an ecosystem, the greater the chances that it will be able to compensate for the changes imposed upon it." See *Environmental Quality* (Washington, DC, 1970), 8, quoted in Oates, *Earth Rising,* 65–66. Paul B. Sears, "Ecology—A Subversive Subject," *BioScience* 14, no. 7 (July 1964): 11; Egler, "Pesticides—In Our Ecosystem," 119, 115, emphasis mine.
34. Robert Disch, ed., *The Ecological Conscience: Values for Survival* (Englewood Cliffs, NJ: Prentice-Hall, 1969), xiv–xv.
35. A. Starker Leopold, "Editor's Foreword," in *The Subversive Science: Essays towards an Ecology of Man,* ed. Paul Shepard and Daniel McKinley (Boston: Houghton Mifflin, 1970), vi.
36. Robert Gottlieb, *Forcing the Spring: The Transformation of the American Environmental Movement* (Washington, DC: Island Press, 2005), 37; Adam Rome, *The Genius of Earth Day: How a 1970 Teach-In Unexpectedly Made the First Green Generation* (New York: Hill & Wang, 2013), x–xi.
37. Rome, *Earth Day,* 9, 30–32; J. Michael McCloskey, *In the Thick of It: My Life in the Sierra Club* (Washington, DC: Island Press, 2005), 106–7.
38. Rome, *Earth Day,* 60, 65, 67–68.
39. Stewart Udall, *The Quiet Crisis* (New York: Holt, Rinehart, and Winston, 1963), 183, 185. In 1967 Galbraith published a follow-up work, *The New Industrial State,* in which he blamed the concentration of economic and political power in large corporations for the problems he had identified in his earlier work. Galbraith's analysis made a point of convergence between the politics of the liberal establishment and that of the New Left. As a result, New Leftists like Ralph Nader, who had been working mostly outside of government circles to demand more constraints on corporate power, joined the confluence of scientists and Great Society liberals headed toward the Earth Day watershed. See Rome, *Earth Day,* 40.
40. Udall, *Quiet Crisis,* vii–viii.
41. The insecticide parathion is a derivative of Nazi nerve gas, Carson informed her readers. Rachel Carson, *Silent Spring* (New York: Fawcett World Library, 1962), 16, 18, 35.
42. Carson, *Silent Spring,* 13–14, 16, 45–50, 53, 58, 83; Udall, *Quiet Crisis,* viii.
43. Rome, *Earth Day,* 24–27; Michael Egan, *Barry Commoner and the Science of Survival: The Remaking of American Environmentalism* (Cambridge, MA: MIT Press, 2007), 127.
44. Barry Commoner, *Science and Survival* (New York: Viking Press, 1966), 80, 83, 125, 151.
45. Paul Ehrlich, *The Population Bomb* (New York: Ballantine, 1968), 47–55, 65; Rome, *Earth Day,* 74. David Brower, as executive director of the Sierra Club

in 1968, had recruited Ehrlich to write the book; see Paul Sabin, *The Bet: Paul Ehrlich, Julian Simon, and Our Gamble Over Earth's Future* (New Haven, CT: Yale University Press, 2013), 10.

46. McCloskey, *In the Thick of It*, 101–2.
47. McCloskey, "Foreword," in *Ecotactics*, ed. Mitchell, 11; John G. Mitchell, "On the Spoor of the Slide Rule," in *Ecotactics*, ed. Mitchell, 23–25.
48. Ralph Nader, "Introduction," in *Ecotactics*, ed. Mitchell, 13. Aldo Leopold had written in "Thinking Like a Mountain," "We all strive for safety, prosperity, comfort, long life . . . but too much safety seems to yield only danger in the long run." Aldo Leopold, *A Sand County Almanac* (New York: Oxford University Press, 1966), 141. Carson, *Silent Spring*, 217–18.
49. Carson, *Silent Spring*, 20, 218, 231, 261.
50. Snyder, *Gary Snyder Reader*, 249; Barry Commoner, *The Closing Circle: Nature, Man, and Technology* (New York: Knopf, 1971), 295.
51. Kenneth Boulding, "The Economics of the Coming Spaceship Earth," in *Environmental Quality in a Growing Economy: Essays from the Sixth RFF Forum*, ed. Henry Jarrett (Baltimore: Johns Hopkins University Press, 1966), 9–10.
52. Mitchell, "On the Spoor," 25, 32, 34–35.
53. Mitchell, "On the Spoor," 24.
54. *Whole Earth Catalog*, Fall 1968, https://monoskop.org/images/0/09/Brand_Stewart_Whole_Earth_Catalog_Fall_1968.pdf; R. Buckminster Fuller, *Operating Manual for Spaceship Earth* (Carbondale: Southern Illinois University Press, 1969), 15–16, 18, 22, 24, 29, http://designsciencelab.com/resources/OperatingManual_BF.pdf.
55. See Roderick Nash, *Wilderness and the American Mind*, 3rd ed. (New Haven, CT: Yale University Press, 1982), 29–31, 161. Ian McHarg, *Design with Nature* (Garden City, NY: Natural History Press, 1969), 7, 13, 35–38 (emphasis mine).
56. Morris Berman, *The Reenchantment of the World* (Ithaca, NY: Cornell University Press, 1981), 295–96.
57. See Rebecca Reider, *Dreaming the Biosphere: The Theater of All Possibilities* (Albuquerque: University of New Mexico Press, 2009), 46–47; and John Allen, Tango Parrish, and Mark Nelson, "The Institute of Ecotechnics," *Environmentalist* 4 (1984): 205–18. Henry Trim, "A Quest for Permanence: The Ecological Visioneering of John Todd and the New Alchemy Institute," in *Groovy Science: Knowledge, Innovation, and American Counterculture*, ed. David Kaiser and W. Patrick McCray (Chicago: University of Chicago Press, 2016), 142, 145, 150.
58. David Armstrong, *A Trumpet to Arms: Alternative Media in America* (Boston: South End Press, 1981), 194–96. "Four Changes" was written with input from Michael McClure, Diane di Prima, Alan Watts, and Stewart Brand. Snyder, *Gary Snyder Reader*, 250, 253, 245; Snyder, *Old Ways*, 65; Snyder, *Real Work*, 27, 147.

59. Ervin Laszlo, *Introduction to Systems Philosophy: Toward a New Paradigm of Contemporary Thought* (New York: Gordon and Breach, 1972), 269, 283, 289; Næss, "The Shallow and the Deep," 95; McHarg, *Design with Nature*, 53.

CHAPTER TWO

1. Steve J. Heims, *John von Neumann and Norbert Wiener: From Mathematics to the Technologies of Life and Death* (Cambridge, MA: MIT Press, 1980), 215–16, 219, 185; Norbert Wiener, *The Human Use of Human Beings: Cybernetics and Society*, 2nd rev. ed. (Boston: Houghton Mifflin, 1954), 24, 33; Steve Joshua Heims, *The Cybernetics Group* (Cambridge, MA: MIT Press, 1991), 12, 15; W. Ross Ashby, *Design for a Brain: The Origin of Adaptive Behaviour*, 2nd rev. ed. (New York: John Wiley & Sons, 1960), v, 37.
2. Heims, *Cybernetics Group*, 283; Wiener, *Human Use*, 38. See also Warren S. McColluch, "Toward Some Circuitry of Ethical Robots or an Observational Science of the Genesis of Social Evaluation in the Mind-Like Behavior of Artifacts," *Acta Biotheoretica* 11 (1956): 147–56, esp. 156; and Norbert Wiener, *God and Golem, Inc.: A Comment on Certain Points Where Cybernetics Impinges on Religion* (Cambridge, MA: MIT Press, 1964), 21–22, 27.
3. Gregory Bateson, *Steps to an Ecology of Mind: Collected Essays in Anthropology, Psychiatry, Evolution, and Epistemology* (Chicago: University of Chicago Press, 1972), 434, 466–67. See also Fritjof Capra, *The Turning Point: Science, Society, and the Rising Culture* (New York: Bantam, 1982), 288.
4. Bateson, *Steps*, 466.
5. Ludwig von Bertalanffy, *General System Theory: Foundations, Development, Applications*, rev. ed. (New York: George Braziller, 1968), 10–12, 14, 16–17, 33, 37; Ludwig von Bertalanffy, "The Role of Systems Theory in Present-Day Science, Technology and Philosophy," in *Toward a Man-Centered Medical Science*, ed. Karl E. Schaefer, Herbert Hensel, and Ronald Brady (Mt. Kisco, NY: Futura, 1977), 11–13.
6. Manfred Drack, "On the Making of a System Theory of Life: Paul A. Weiss and Ludwig von Bertalanffy's Conceptual Connection," *Quarterly Review of Biology* 82, no. 4 (December 2007): 362.
7. Ervin Laszlo, *Introduction to Systems Philosophy: Toward a New Paradigm of Contemporary Thought* (New York: Gordon and Breach, 1972), viii–ix. See Ervin Laszlo, *Simply Genius! And Other Tales from My Life: An Informal Autobiography* (Carlsbad, CA: Hay House, 2011), 19–20, 26, 54, 128–30.
8. In the same year that *The Systems View of the World* appeared, Laszlo also published *Introduction to Systems Philosophy* and *The Relevance of General Systems Theory: Papers Presented to Ludwig von Bertalanffy on His Seventieth Birthday* (New York: George Braziller, 1972).
9. Ervin Laszlo, *The Systems View of the World* (New York: George Braziller, 1972), 80; Dorothy Emmett, "Alfred North Whitehead," in *The Encyclopedia of Philosophy*, vol. 8, ed. Paul Edwards (New York: Macmillan and the

Free Press, 1967), 290; Daniel Belgrad, *The Culture of Spontaneity: Improvisation and the Arts in Postwar America* (Chicago: University of Chicago Press, 1998), 125–27. See also T. F. H. Allen and Thomas Hoekstra, *Toward a Unified Ecology* (New York: Columbia University Press, 1992), 25.

10. Laszlo, *Systems View*, 20, 81.
11. Laszlo, *Systems View*, 8–9.
12. Laszlo, *Systems View*, 33, 51, 115.
13. Laszlo, *Introduction to Systems*, 51.
14. Laszlo, *Systems View*, 46.
15. Laszlo, *Systems View*, 67, 37.
16. See Allen and Hoekstra, *Toward a Unified Ecology*, 32–33.
17. Laszlo, *Systems View*, 113, 118.
18. Laszlo, *Introduction to Systems*, 171, 241; Laszlo, *Systems View*, 115, 113.
19. Paul A. Weiss, *The Science of Life: The Living System—A System for Living* (Mt. Kisco, NY: Futura, 1973), 52. See also Bertalanffy, "The Role of Systems Theory," 13; and Paul A. Weiss, "The System of Nature and the Nature of Systems," in *Toward a Man-Centered Medical Science*, ed. Schaefer, Hensel, and Brady, 29, 37.
20. Laszlo, *Systems View*, 52, 109–10, 113.
21. Laszlo, *Systems View*, 115, 117, 103, 120.
22. Laszlo, *Systems View*, 10–11, 15, 79, 87, 89–90.
23. Laszlo, *Systems View*, 65, 108, 118; Laszlo, *Introduction to Systems*, 269, 283, 289.
24. Sharon E. Kingsland, *Modeling Nature: Episodes in the History of Population Ecology* (Chicago: University of Chicago Press, 1985), 1.
25. Arthur Koestler and J. R. Smythies, eds., *Beyond Reductionism: New Perspectives in the Life Sciences* (Boston: Beacon Press, 1969), v.
26. In ways that I have not yet been able to trace, this idea seems to have influenced American racial politics of the seventies. Following the original integrationist impulse of the civil rights movement and the subsequent separatist reaction resulting in Black, Chicano, and Native American nationalisms, the embrace of racial diversity and "multiculturalism" in the seventies marked a third phase of the struggle, consistent with an image of American society as a network of semi-discrete but interfaced cultural subsystems. In the dynamics defined by multiculturalism, ethnic identities are Janus-faced subcultures: "holons," in Arthur Koestler's terminology. They face outward toward a cultural environment constituted by all other ethnicities and their interactions, while internally maintaining the features particular to their own cultural community. Literary works of the seventies explored this cultural terrain, manifesting ethnicity's Janus-faced quality in their use of "code switching": a movement back and forth between two languages and styles, representing an interface along the permeable membrane of a hyphenated American identity that was by turns "ethnic" and "mainstream." Examples of such literature

include Rudolfo Anaya's *Bless Me, Ultima* (1972), Maxine Hong Kingston's *The Woman Warrior* (1976), and Leslie Marmon Silko's *Ceremony* (1977). See Bruce J. Schulman, *The Seventies: The Great Shift in American Culture, Society, and Politics* (New York: Free Press, 2001), 58–72; Paul Gilroy, *The Black Atlantic: Modernity and Double Consciousness* (Cambridge, MA: Harvard University Press, 1993), 122–24; and Daniel Belgrad, "Performing *lo Chicano*," *MELUS* 29, no. 2 (Summer 2004): 251.

27. Laszlo, *Introduction to Systems*, 48, 51; Erich Jantsch, *The Self-Organizing Universe: Scientific and Human Implications of the Emerging Paradigm of Evolution* (New York: Pergamon Press, 1980), 68.
28. Douglas Robertson, "Feedback Theory and Darwinian Evolution," *Journal of Theoretical Biology* 152, no. 4 (October 21, 1991): 470.
29. Conrad Waddington, "The Theory of Evolution Today," in *Beyond Reductionism*, ed. Koestler and Smythies, 364.
30. Waddington, "The Theory of Evolution Today," 364, 374, 371. See also Gregory Bateson, *Mind and Nature: A Necessary Unity* (New York: Dutton, 1979), 156. James Mark Baldwin's thinking influenced Bateson and multiple attendees at Koestler's Alpbach conference, including Jean Piaget, Paul Weiss, and Waddington. For more on Baldwin and his influence, see Robert Richards, *Darwin and the Emergence of Evolutionary Theories of Mind and Behavior* (Chicago: University of Chicago Press, 1987), 399; R. H. Wozniak, "Consciousness, Social Heredity, and Development: The Evolutionary Thought of James Mark Baldwin," *American Psychologist* 64 (2009): 93–101; J. T. Burman, "Updating the Baldwin Effect: The Biological Levels Behind Piaget's New Theory," *New Ideas in Psychology* 31, no. 3 (2013): 363–73; and Brian Cox, "Editorial: The Past and Future of Epigenesis in Psychology," *New Ideas in Psychology* 31, no. 3 (2013): 351–54.
31. Paul Ehrlich and Peter Raven, "Butterflies and Plants: A Study in Coevolution," *Evolution* 18, no. 4 (1964): 586–608.
32. Paul Shepard, *Thinking Animals: Animals and the Development of Human Intelligence* (New York: Viking, 1978), 10, 15.
33. Bateson, *Steps*, 438, 313, 319.
34. Bateson, *Steps*, 320. Bateson's ideas influenced Marshall McLuhan, who wrote in *Understanding Media* (1964) that communications media functioned as a global nervous system, an extension of the individual nervous system and its circulation of signals. See Fred Turner, *From Counterculture to Cyberculture: Stewart Brand, the Whole Earth Network, and the Rise of Digital Utopianism* (Chicago: University of Chicago Press, 2006), 53. See also Andrew Pickering, *The Cybernetic Brain: Sketches of Another Future* (Chicago: University of Chicago Press, 2010), 21–23, 30.
35. Weiss, "The System of Nature and the Nature of Systems," 31.
36. Ludwig von Bertalanffy, "Chance or Law," in *Beyond Reductionism*, ed. Koestler and Smythies, 69–70. The mathematical term for this is "self-similarity."

37. Scott MacDonald, "Godfrey Reggio," in *A Critical Cinema 2: Interviews with Independent Filmmakers* (Berkeley: University of California Press, 1992), 378–79; Victor Rivas López, "A Phenomenological Approach to Earth Oblivion and Human Unbalance in *Koyaanisqatsi*," in *The Cosmos and the Creative Imagination*, ed. Anna-Teresa Tymieniecka and Patricia Trutty-Coohill (New York: Springer International, 2016), 364.
38. Turner, *From Counterculture to Cyberculture*, 43; David Lipset, *Gregory Bateson: The Legacy of a Scientist* (Englewood Cliffs, NJ: Prentice-Hall, 1980), 285–86; Stewart Brand, "Review of *Steps to an Ecology of Mind*," *Rolling Stone*, November 9, 1972, 76; Stewart Brand, "Both Sides of the Necessary Paradox," *Harper's*, November 1973, 20–37. Another version of their conversations appeared as "Both Sides of the Necessary Paradox (Conversations with Gregory Bateson)" in Brand's book *II Cybernetic Frontiers* (New York: Random House, 1974).
39. Lynn Margulis and James Lovelock, "The Atmosphere as Circulatory System of the Biosphere—the Gaia Hypothesis," reprinted in *News That Stayed News: Ten Years of CoEvolution Quarterly, 1974–1984*, ed. Art Kleiner and Stewart Brand (San Francisco: North Point Press, 1986), 15–25; Dian R. Hitchcock and James E. Lovelock, "Life Detection by Atmospheric Analysis," *Icarus* 7, no. 2 (September 1967): 149–59, http://jameslovelock.org/page28.html; this argument was developed in the context of having being asked by NASA to find ways of remotely detecting the presence of life on other planets. James E. Lovelock, "Gaia as Seen through the Atmosphere," *Atmospheric Environment* 6, no. 8 (August 1972): 579–80; James Lovelock, *Gaia: A New Look at Life on Earth* (New York: Oxford University Press, 1979), 2, 6, 10.
40. Bruce Clarke, "Gaia Is Not an Organism," in *Lynn Margulis: The Life and Legacy of a Scientific Rebel*, ed. Dorion Sagan (White River Junction, VT: Chelsea Green, 2012), 37–38; Lynn Margulis and James E. Lovelock, "Biological Modulation of the Earth's Atmosphere," *Icarus* 21, no. 4 (April 1974): 471–89; James E. Lovelock and Lynn Margulis, "Atmospheric Homeostasis by and for the Biosphere: The Gaia Hypothesis," *Tellus* 26, nos. 1–2 (February 1974): 2–10; James E. Lovelock and Lynn Margulis, "Homeostatic Tendencies of the Earth's Atmosphere," *Origins of Life* 5 (1974): 93–103.
41. James Lovelock and Sidney Epton, "The Quest for Gaia," *New Scientist* 65, no. 935 (February 6, 1975): 304; Lynn Margulis, *Symbiotic Planet: A New Look at Evolution* (New York: Basic Books, 1998), 119, 123; Lovelock, *Gaia*, 1; Capra, *Turning Point*, 285.
42. Morris Berman, *The Reenchantment of the World* (Ithaca, NY: Cornell University Press, 1981), 259.
43. Lovelock, *Gaia*, 8, 15–16, 28, 31, 41, 82, 107–10, 112, 115, 120.
44. Lovelock, *Gaia*, 147.
45. Turner, *From Counterculture to Cyberculture*, 126; Andrew G. Kirk, *Counterculture Green: The Whole Earth Catalog and American Environmentalism* (Lawrence: University Press of Kansas, 2007), 211.

46. Turner, *From Counterculture to Cyberculture*, 126; Gary Snyder, *The Real Work: Interviews and Talks, 1964–1979* (New York: New Directions, 1980), 126.
47. Snyder, *Real Work*, 82.
48. Gary Snyder, *The Gary Snyder Reader* (Washington, DC: Counterpoint, 1999), 250, 252; Snyder, quoted in Pat Smith and Mariana Gosnell, "That Snyder Sutra," in *Ecotactics: The Sierra Club Handbook for Environmental Activists*, ed. John Mitchell (New York: Pocket Books, 1970), 86; Snyder, *Real Work*, 130.
49. Leonard Scigaj, *Sustainable Poetry* (Lexington: University of Kentucky Press, 1999), 232–34; Snyder, *Real Work*, 44–50; Gary Snyder, *The Old Ways: Six Essays* (San Francisco: City Lights Books, 1977), 10, 12–13, 18–19. This was at a time when Congress was debating some provisions of the Marine Mammal Protection Act of 1972 and the Endangered Species Act of 1973. See Rep. Gerry Studds, "Introduction of Legislation to Save Scrimshaw Industry," in U.S. Congress, *Congressional Record: Proceedings and Debates of the 93rd Congress, Second Session*, vol. 120, part 17, page 23011.

CHAPTER THREE

1. Gary Snyder, *The Real Work: Interviews and Talks, 1964–1979* (New York: New Directions, 1980), 130, 28, 69; Gary Snyder, *The Old Ways: Six Essays* (San Francisco: City Lights Books, 1977), 27–29 (emphases mine).
2. Eugenie Brinkema, *The Forms of the Affects* (Durham, NC: Duke University Press, 2014), 4.
3. Brinkema, *Forms of the Affects*, 6–8.
4. Stewart Udall, *The Quiet Crisis* (New York: Holt, Rinehart, and Winston, 1963), 4–5, 12, 177.
5. Theodore Roszak, *The Making of a Counterculture: Reflections on the Technocratic Society and Its Youthful Opposition* (New York: Anchor Books, 1969), 51–52.
6. Gregory Bateson, *Steps to an Ecology of Mind: Collected Essays in Anthropology, Psychiatry, Evolution, and Epistemology* (Chicago: University of Chicago Press, 1972), xxvii, xxxii, xxix.
7. Stewart Brand, "Both Sides of the Necessary Paradox," *Harper's*, November 1973, 32, 34.
8. See Bernard Williams, "Rene Descartes," in *The Encyclopedia of Philosophy*, vol. 2, ed. Paul Edwards (New York: Macmillan and the Free Press, 1967), 351–54.
9. Ervin Laszlo, *Introduction to Systems Philosophy: Toward a New Paradigm of Contemporary Thought* (New York: Gordon and Breach, 1972), 221; Dorothy Emmett, "Alfred North Whitehead," in *The Encyclopedia of Philosophy*, vol. 8, ed. Edwards, 293.
10. Morris Berman, *The Reenchantment of the World* (Ithaca, NY: Cornell University Press, 1981), 15, 250–54, 274. Berman also cites Whitehead's

process philosophy, as well as Wilhelm Reich's psychology, gestalt psychology, and phenomenology, as alternatives to the Cartesian dualism (135, 156, 137, 145, 147).

11. Fritjof Capra, *The Turning Point: Science, Society, and the Rising Culture* (New York: Bantam, 1982), 41.
12. Snyder, *Old Ways*, 37.
13. Elizabeth Dodson Gray, *Green Paradise Lost* (Wellesley, MA: Roundtable Press, 1979), 13, 100, 25.
14. Margot Adler, *Drawing Down the Moon: Witches, Druids, Goddess-Worshippers, and Other Pagans in America Today*, rev. and exp. ed. (Boston: Beacon Press, 1986), 22–23; Gray, *Green Paradise Lost*, 82. There were also theologians who attempted to reconcile Christianity with an ecological ethic; see, for instance, H. Paul Santmire, *Brother Earth: Nature, God and Ecology in Time of Crisis* (New York: Thomas Nelson, 1970); and Charles Birch and John B. Cobb Jr., *The Liberation of Life: From the Cell to the Community* (New York: Cambridge University Press, 1981).
15. Lynn White Jr., "The Historical Roots of Our Ecologic Crisis," in *Western Man and Environmental Ethics: Attitudes toward Nature and Technology*, ed. Ian Barbour (Reading, MA: Addison-Wesley, 1973), 24–25, 27–28. In 1979 Pope John Paul II proclaimed Saint Francis the patron saint of ecology.
16. Lynn White Jr., "The Historical Roots of Our Ecologic Crisis," *Science* 155, no. 3767 (March 10, 1967): 1203–7; Garrett de Bell, ed., *Environmental Handbook: Prepared for the First National Environmental Teach-In* (New York: Ballantine, 1970); Gray, *Green Paradise Lost*, viii.
17. Gray, *Green Paradise Lost*, 3, 46.
18. Gray, *Green Paradise Lost*, 87–89, 38, 130. See also Michael Rogin, "Liberal Society and the Indian Question," in *"Ronald Reagan," the Movie: And Other Episodes in Political Demonology* (Berkeley: University of California Press, 1987).
19. Gray, *Green Paradise Lost*, 125. See also Rosemary Radford Ruether, *New Woman, New Earth: Sexist Ideologies and Human Liberation* (New York: Seabury Press, 1975), 186–204.
20. Gray, *Green Paradise Lost*, 61, 64–65, 67, 83, 97.
21. Gray, *Green Paradise Lost*, 82, 111, 114–16.
22. Gray, *Green Paradise Lost*, 58, 71, 76, 78.
23. Gray, *Green Paradise Lost*, 55, 85; Gary Snyder, "Energy Is Eternal Delight," *New York Times*, January 12, 1972, 43; reprinted in Snyder, *Turtle Island* (New York: New Directions, 1974), 105.
24. Philip J. Deloria, *Playing Indian* (New Haven, CT: Yale University Press, 1998), 156, 169.
25. See Brewton Berry, "The Myth of the Vanishing Indian," *Phylon* 21, no. 1 (1960): 51–57. Other widely recognized examples of the crying Indian from seventies popular culture are Will Sampson's portrayal of Chief Bromden in the 1975 film *One Flew Over the Cuckoo's Nest* and Iron Eyes

Cody as Standing Bear in the revisionist Western *Grayeagle* (1977). In the former, Chief Bromden cries when he euthanizes his buddy Randle McMurphy (played by Jack Nicholson), who has been lobotomized by the establishment for bucking institutional injustices. The scene is accompanied by "Indian" music (drum, rattle, and flute) that returns when Chief Bromden escapes into the natural landscape. In the latter, Standing Bear helps his best friend, the white settler John Coulter, track a Cheyenne Indian (Grayeagle) who has kidnapped Coulter's daughter. But a plot twist reveals that Grayeagle is taking her to see her true biological father, the Indian chief Running Wolf, who is now old and dying but who in his youth had had an affair with Coulter's wife—something that Standing Bear knew all along. Standing Bear cries when Running Wolf dies, symbolizing the passing of the unassimilated Indian.

26. Amy Waldman, "Iron Eyes Cody, 94, an Actor and Tearful Anti-Littering Icon," *New York Times*, January 5, 1999, http://www.nytimes.com/1999/01/05/arts/iron-eyes-cody-94-an-actor-and-tearful-anti-littering-icon.html.
27. Carlos Castaneda, *The Teachings of Don Juan: A Yaqui Way of Knowledge* (Berkeley: University of California Press, 1998), 2, 5–7; "Don Juan and the Sorcerer's Apprentice," *Time*, March 5, 1973; Richard de Mille, *The Don Juan Papers: Further Castaneda Controversy* (Santa Barbara, CA: Ross-Erikson, 1980).
28. Gerald Robert Vizenor, *Fugitive Poses: Native American Indian Scenes of Absence and Presence* (Lincoln: University of Nebraska Press, 2000), 67–69; Alex Jacobs, "Fool's Gold: The Story of Jamake Highwater, the Fake Indian Who Won't Die," *Indian Country Today*, June 19, 2015, https://indiancountrymedianetwork.com/culture/arts-entertainment/fools-gold-the-story-of-jamake-highwater-the-fake-indian-who-wont-die/; Jack Anderson, "A Fabricated Indian?" *Washington Post*, February 16, 1984; Hank Adams, "The Golden Indian," *Akwesasne Notes* 16 (Summer 1984): 10–12.
29. Alice Kehoe, "Primal Gaia: Primitivists and Plastic Medicine Men," in *The Invented Indian: Cultural Fictions and Government Policies*, ed. James Clifton (New Brunswick, NJ: Transaction, 1990), 193–210; Daniel Belgrad, *The Culture of Spontaneity: Improvisation and the Arts in Postwar America* (Chicago: University of Chicago Press, 1998), 113.
30. Tom Brown, *The Tracker: The Story of Tom Brown, Jr.*, as told to William Jon Watkins (Englewood Cliffs, NJ: Prentice-Hall, 1978), 17–19, 31–32, 140.
31. Tom Brown, "Ancestors," in *Deep Ecology*, ed. Michael Tobias, 2nd ed. (San Diego: Avant Books, 1988), 125, 128, 137–38.
32. Brown, *Tracker*, 19.
33. Andrew Pickering, *The Cybernetic Brain: Sketches of Another Future* (Chicago: University of Chicago Press, 2010), 21–25, 30; Brown, *Tracker*, 25.
34. Farley Mowat, *Never Cry Wolf* (New York: Bantam, 1973), 60.
35. Brown, *Tracker*, 29.
36. Brown, *Tracker*, 25, 27.

37. Snyder, *Real Work*, 34–36.
38. Brown, *Tracker*, 26–27.
39. Brown, *Tracker*, 27, 89, 35.
40. Mowat, *Never Cry Wolf*, v. The informed reader is therefore left asking whether Ootek ever existed, any more than Castaneda's Don Juan. By comparing Mowat's various versions of events, one comes to the conclusion that there was an actual Ootek, but that the character "Ootek" is a composite of several men whom Mowat met in the North country: a distillation of what he perceived to be the Inuit's best qualities, just as "Mike" is a composite of attitudes exhibited by some of those same men, invented to create a foil for "Ootek." See Mowat, *Never Cry Wolf*, 72, 83, 89–91, 121–22, 128–34; Farley Mowat, *People of the Deer*, rev. ed. (New York: Carroll & Graf, 1975), 116, 155, 221, 232–35; and Farley Mowat, *The Desperate People*, rev. ed. (Toronto: McClelland & Stewart, 1975), 22, 27–29, 44, 47, 127.
41. Mowat, *People of the Deer*, 306. For a summary of the debate over Mowat's veracity that concludes favorably for Mowat, see T. Querengesser, "Farley Mowat: Liar or Saint?," *Up Here*, September 2009, https://web.archive.org/web/20130220124959/http://www.uphere.ca/node/442.
42. For a discussion of the concept of "middle ground," see Richard White, *Middle Ground: Indians, Empires, and Republics in the Great Lakes Region, 1650–1815* (New York: Cambridge University Press, 1991), x. Another such Indian intellectual was N. Scott Momaday, professor of literature at the University of California, Berkeley, whose novel *House Made of Dawn* won the 1969 Pulitzer Prize for fiction. Momaday wrote an essay for inclusion in the Sierra Club's Earth Day publication *Ecotactics* on the relevance of the Native American land ethic. See N. Scott Momaday, "An American Land Ethic," in *Ecotactics: The Sierra Club Handbook for Environmental Activists*, ed. John Mitchell (New York: Pocket Books, 1970), 97–105.
43. Akwesasne Notes, ed., *A Basic Call to Consciousness* (Rooseveltown, NY: Mohawk Nation, 1978), 77. The newspaper was founded in 1968 by Jerry Gambill (Rarihokwats), a non-Indian community development worker with the Canadian Department of Indian Affairs. John Mohawk, a member of the Seneca tribe of western New York, took over as editor in 1973. See Laurence M. Hauptman, *Seven Generations of Iroquois Leadership: The Six Nations since 1800* (Syracuse, NY: Syracuse University Press, 2008), 179–80.
44. Akwesasne Notes, *Basic Call to Consciousness*, 80.
45. Akwesasne Notes, *Basic Call to Consciousness*, 80, 71–72; Gayle High Pine, "The Non-Progressive Great Spirit," *Akwesasne Notes* 5, no. 6 (Early Winter 1973): 38, quoted in Adler, *Drawing Down the Moon*, 381.
46. Akwesasne Notes, *Basic Call to Consciousness*, 73–77. Friedrich Engels had made a similar argument in *The Origin of the Family, Private Property, and the State* (1884).
47. Akwesasne Notes, *Basic Call to Consciousness*, 78–79.

48. Vine Deloria Jr., *God Is Red: A Native View of Religion* (New York: Dell, 1973), 71. Deloria was among those who helped to expose Jamake Highwater as a fake Indian in the pages of the *Akwesasne Notes*. See Jacobs, "Fool's Gold."
49. Bateson, *Steps*, 468, quoted in Vine Deloria Jr., *The Metaphysics of Modern Existence* (New York: Harper & Row, 1979), 20.
50. Deloria, *Metaphysics*, 154.
51. Deloria, *Metaphysics*, viii, xiii, 17–18, 34, 37, 160.
52. Deloria, *Metaphysics*, 196–201, 82; Frank A. Brown Jr., "The Rhythmic Nature of Animals and Plants," *American Scientist* 47, no. 2 (June 1959): 160–61, 164, 166, 168.
53. Deloria, *Metaphysics*, viii–ix, 161.

CHAPTER FOUR

1. "Vegetable" and "Vegetal," *Compact Edition of the Oxford English Dictionary* (New York: Oxford University Press, 1971), 2:3598.
2. Rachel Carson, *Silent Spring* (New York: Fawcett World Library, 1962), 64; Adam Rome, *The Genius of Earth Day: How a 1970 Teach-In Unexpectedly Made the First Green Generation* (New York: Hill & Wang, 2013), 43; Kenneth Cauthen, *Process Ethics: A Constructive System* (New York: E. Mellen Press, 1984), 72.
3. Christopher D. Stone, "Should Trees Have Standing? Toward Legal Rights for Natural Objects," *Southern California Law Review* 45 (1972): 456. Rogers Morton was Nixon's secretary of interior who approved the plan on behalf of the Forest Service.
4. Stone, "Should Trees Have Standing?" 498.
5. Stone, "Should Trees Have Standing?" 453–54, 471, 498–99.
6. Stone, "Should Trees Have Standing?" 478–79.
7. Stone, "Should Trees Have Standing?" 496, 498.
8. *Sierra Club v. Morton*, 405 U.S. 727 at 743 (1972) (Douglas, J. *dissenting*), https://supreme.justia.com/cases/federal/us/405/727/case.html.
9. *Sierra Club v. Morton*, 743; Gary Snyder, *The Old Ways: Six Essays* (San Francisco: City Lights Books, 1977), 13; Stone, "Should Trees Have Standing?" 471.
10. Other contenders for inventing the lie detector machine include Leonarde Keeler, who called it the "emotograph," and John Larson, who called it the "cardio-pneumo psychogram."
11. Jon Atack, *A Piece of Blue Sky: Scientology, Dianetics and L. Ron Hubbard Exposed* (New York: Carol, 1990), 22. When Hubbard was threatened with legal action for practicing medicine without a license, he reinvented Dianetics as a religion.
12. Subrata Dasgupta, "Jagadis Bose, Augustus Waller and the Discovery of 'Vegetable Electricity,'" *Notes and Records of the Royal Society of London*

52, no. 2 (July 1998): 313–14; "Dr. Hubbard Seen by TV Millions," *Garden News* (London), December 18, 1959, 21, http://www.lisamcpherson.org/news/1953-59/newspaper_1953-59.pdf.

13. Cleve Backster, "Evidence of Primary Perception in Plant Life," *International Journal of Parapsychology* 10 (1968): 329–48; Thorn Bacon, "The Man Who Reads Nature's Secret Signals," *National Wildlife* 7 (February–March 1969), 8.
14. Richard Martin, "Be Kind to Plants—Or You Could Cause a Violet to Shrink," *Wall Street Journal*, February 2, 1972, 1, 17; "Talk Nice to Plants . . . They May Be Listening!" *Press Democrat* (Santa Rosa, CA), September 31, 1972, www.ebdir.net/enlighten/; Christopher Bird and Peter Tompkins, "Love among the Cabbages," *Harper's*, November 1972, 90; Peter Tompkins and Christopher Bird, *The Secret Life of Plants* (New York: Harper, 1973), 8–9, 11.
15. Adolph Hecht, "Emotional Responses by Plants," *Plant Science Bulletin* 20, no. 4 (December 1974): 46–47.
16. Cleve Backster, *Primary Perception: Biocommunication with Plants, Living Foods, and Human Cells* (Anza, CA: White Rose Millennium Press, 2003), 17, 23.
17. Tompkins and Bird, *Secret Life of Plants*, 6, xi. They also wrote that since "in plants there [is] respiration without gills or lungs, digestion without a stomach, and movements without muscles," it is plausible that they could also have feelings without a central nervous system (92). Similar thinking led French philosophers Gilles Deleuze and Félix Guattari in 1976 to use the plant rhizome as a trope to describe their ideal of a nonhierarchical, a-centered subjectivity, which they referred to as the "body without organs." See Gilles Deleuze and Félix Guattari, *Rhizome: Introduction* (Paris: Éditions de Minuit, 1976).
18. Backster, *Primary Perception*, 30–31; Tompkins and Bird, *Secret Life of Plants*, 9.
19. Tompkins and Bird, *Secret Life of Plants*, xii, 12–14; Backster, *Primary Perception*, 29–30, 33–34; *The Secret Life of Plants*, dir. Walon Green (Infinite Enterprises, 1978).
20. Tompkins and Bird, *Secret Life of Plants*, 7–8. On the phenomenon of threat-induced fainting among humans, see H. Stefan Bracha, "Freeze, Flight, Fight, Fright, Faint," *CNS Spectrums* 9, no. 9 (September 2004): 679–85.
21. Kenneth Horowitz, Donald Lewis, and Eegar Gasteiger, "Plant 'Primary Perception': Electrophysiological Unresponsiveness to Brine Shrimp Killing," *Science* 189, no. 4201 (August 8, 1975): 478–80; John M. Kmetz, "A Study of Primary Perception in Plants and Animal Life," *Journal of the American Society for Psychical Research* 71, no. 2 (1977): 157–70.
22. Backster, *Primary Perception*, 29, 35, 49, 124; Tompkins and Bird, *Secret Life of Plants*, 7; Gregory Bateson, *Steps to an Ecology of Mind: Collected Essays in*

Anthropology, Psychiatry, Evolution, and Epistemology (Chicago: University of Chicago Press, 1972), 368.

23. Tompkins and Bird, *Secret Life of Plants*, 27.
24. Backster, *Primary Perception*, 12; Tompkins and Bird, *Secret Life of Plants*, 19, 22–23.
25. Ivan Sharpe, "E. Bay Pair Prove That Plants Lead Secret Lives," *San Francisco Examiner*, March 28, 1977, http://www.ebdir.net/enlighten/news_sf_exam_1977_03_28.html; Tompkins and Bird, *Secret Life of Plants*, 32.
26. Backster, *Primary Perception*, 114–15; H. E. Puthoff and R. Fontes, "Organic Biofield Sensor," Stanford Research Institute Electronic and Bioengineering Laboratory, Menlo Park, CA, November 1975, 1, 5.
27. Puthoff and Fontes, "Organic Biofield Sensor," 33; Sharpe, "E. Bay Pair." See also Glennda Chui, "Castro Valley Men Probe the Secret Life of Plants," *Daily Review* (Hayward, CA), August 5, 1977, http://www.ebdir.net/enlighten/news_daily_review_1977_08_05.html.
28. Tompkins and Bird, *Secret Life of Plants*, 28–29, 31–32. For more on the influence of Reichian psychology in American popular culture, see Daniel Belgrad, *The Culture of Spontaneity: Improvisation and the Arts in Postwar America* (Chicago: University of Chicago Press, 1998), 159, 151; and Richard King, *The Party of Eros: Radical Social Thought and the Realm of Freedom* (Chapel Hill: University of North Carolina Press, 1972). See also Dolores LaChapelle, "Sacred Land, Sacred Sex," in *Deep Ecology*, ed. Michael Tobias (San Diego: Avant Books, 1988), 117.
29. Snyder, *Old Ways*, 9, 12; Michael Marder and Luce Irigaray, *Through Vegetal Being: Two Philosophical Perspectives* (New York: Columbia University Press, 2016), 158.
30. Barbara B. Brown, *New Mind, New Body: Bio-Feedback: New Directions for the Mind* (New York: Harper & Row, 1974), 205; Stanley Finger, *Origins of Neuroscience: A History of Explorations into Brain Function* (New York: Oxford University Press, 1994), 282; Neal Miller, "Learning of Visceral and Glandular Responses," *Science* 163 (1969): 434.
31. Backster, *Primary Perception*, 76. Electroencephalography was invented by German neurophysiologist Hans Berger in the 1920s. Berger had originally invented it to pursue research into telepathy and psychophysiology, but he did not realize any applications in that area. See David Millett, "Hans Berger: From Psychic Energy to the EEG," *Perspectives in Biology and Medicine* 44, no. 4 (Autumn 2001): 522–42; and Thomas Collura, "History and Evolution of Electroencephalographic Instruments and Techniques," *Journal of Clinical Neurophysiology* 10, no. 4 (1993): 484, 489, 494.
32. Jerome Bruner, "On Voluntary Action and Its Hierarchical Structure," in *Beyond Reductionism: New Perspectives in the Life Sciences*, ed. Arthur Koestler and J. R. Smythies (Boston: Beacon Press, 1969), 161–63; Brown, *New Mind, New Body*, 32, 64; Andrew Weil, *The Natural Mind: A New Way of Looking at Drugs and the Higher Consciousness* (Boston: Houghton Mifflin,

1972), 161, 163; Fritjof Capra, *The Turning Point: Science, Society, and the Rising Culture* (New York: Bantam, 1982), 351; Joe Kamiya, "Operant Control of the EEG Alpha Rhythm and Some of Its Reported Effects on Consciousness," in *Altered States of Consciousness*, ed. Charles T. Tart, 2nd ed. (New York: Anchor Books, 1972), 520–27.

33. George Fuller, *Biofeedback: Methods and Procedures in Clinical Practice* (San Francisco: Biofeedback Institute of San Francisco, 1977), 60; Brown, *New Mind, New Body*, 49–51.
34. Brown, *New Mind, New Body*, 79–80, 86.
35. Weil, *Natural Mind*, 165–66, 160–61; Andrew Weil, *The Marriage of the Sun and Moon: A Quest for Unity in Consciousness* (Boston: Houghton Mifflin, 1980), 12, 37–39; Andrew Weil and Peter Fremont, "Interview with Andrew Weil," *High Times* 1, no. 5 (August–September 1975): 17.
36. Weil, *Marriage of the Sun and Moon*, 115, 233–34; Weil, *Natural Mind*, 143.
37. Weil, *Natural Mind*, 140, 142; Weil and Fremont, "Interview," 17, 58; Weil, *Marriage of the Sun and Moon*, 12.
38. Weil and Fremont, "Interview," 16.
39. Weil, *Natural Mind*, 132–33.
40. Weil and Fremont, "Interview," 58; Weil, *Marriage of the Sun and Moon*, 3.
41. Gordon Wasson was the first Westerner to ingest *Psilocybe*, in Huautla de Jimenéz in the state of Oaxaca in 1955. See R. Gordon Wasson, Albert Hofmann, and Carl Ruck, *The Road to Eleusis: Unveiling the Secret of the Mysteries* (New York: Harcourt Brace Jovanovich, 1978), 4.
42. Brown, *New Mind, New Body*, 43; Lynn Sagan, "An Open Letter to Mr. Joe K. Adams," *Psychedelic Review* 1, no. 3 (1964): 354; Sharpe, "E. Bay Pair"; Bateson, *Steps*, 469; Simone Forti, *Handbook in Motion* (Northampton, MA: Contact Editions, 1974), 18; Terence McKenna, *True Hallucinations: Being an Account of the Author's Extraordinary Adventures in the Devil's Paradise* (San Francisco: HarperSanFrancisco, 1993), 40.
43. Dennis McKenna, *The Brotherhood of the Screaming Abyss: My Life with Terence McKenna* (St. Cloud, MN: North Star Press, 2012), 246–47, 253, 290–94; McKenna, *True Hallucinations*, 43.
44. O. T. Oss and O. N. Oeric, *Psilocybin: Magic Mushroom Grower's Guide: A Handbook for Psilocybin Enthusiasts* (Berkeley, CA: And/Or Press, 1976), 8–9. Fungal mycelia form underground networks connecting the roots of plants and allowing them to communicate via what Suzanne Simard, forest ecologist, has called "the wood-wide web." See Nic Fleming, "Plants Talk to Each Other Using an Internet of Fungus," BBC, November 11, 2014, http://www.bbc.com/earth/story/20141111-plants-have-a-hidden-internet; and Robert MacFarlane, "The Secrets of the Wood Wide Web," *New Yorker*, August 7, 2016, http://www.newyorker.com/tech/elements/the-secrets-of-the-wood-wide-web.
45. McKenna, *True Hallucinations*, xii, 71–72, 98–99, 108, 209; McKenna, *Brotherhood*, 246.

46. McKenna, *True Hallucinations*, 197, 110–11, 98–99; McKenna, *Brotherhood*, 310–11.
47. McKenna, *True Hallucinations*, 110; Weil, *Marriage of the Sun and Moon*, 60, 77. At the 1968 Alpbach conference organized by Arthur Koestler, biologist Paul Weiss described "field effects" like murmuration and group mind as phenomena of biological systems; see Paul A. Weiss, "The Living System: Determinism Stratified," in *Beyond Reductionism*, ed. Koestler and Smythies, 41. See also Marder and Irigaray, *Through Vegetal Being*, 112.
48. McKenna, *Brotherhood*, 267, 278, 298; McKenna, *True Hallucinations*, 138, 216; Weil, *Marriage of the Sun and Moon*, 101, 195; Weil, *Natural Mind*, 182, 186–87.
49. Stephen Gaskin, *Haight Ashbury Flashbacks* (Berkeley: Ronin, 1990), 51, 56, 109.
50. Gaskin, *Haight Ashbury Flashbacks*, 77, 165, 114, 122.
51. Stephen Diamond, *What the Trees Said: Life on a New Age Farm* (New York: Delacorte Press, 1971), 85.
52. John Jervis, "Uncanny Presences," in *Uncanny Modernity: Cultural Theories, Modern Anxieties*, ed. Jo Collins and John Jervis (New York: Palgrave Macmillan, 2008), 10.
53. Sigmund Freud, *The Uncanny*, trans. by David McLintock (New York: Penguin, 2003), 143, 153; Forbes Morlock, "Doubly Uncanny: An Introduction to 'On the Psychology of the Uncanny,'" *Angelaki* 2, no. 1 (1997): 17–18; Jervis, "Uncanny Presences," 26.
54. Anthony Wilden, *System and Structure: Essays in Communication and Exchange*, 2nd ed. (New York: Tavistock, 1980), xlii, 463–64; Jervis, "Uncanny Presences," 14, 20, 25; Ernst Jentsch, "On the Psychology of the Uncanny," trans. Roy Sellars, in *Uncanny Modernity*, ed. Collins and Jervis, 227.
55. On the far side of this threshold, one is no longer a coherent subject and can only express oneself inarticulately, through bodily flailing and animalistic vocalizations: "irruptions" of the "chora" that Kristeva hypothesized are the basis of all dance and song. Julia Kristeva, *The Powers of Horror: An Essay on Abjection*, trans. Leon Roudiez (New York: Columbia University Press, 1982), 2; Julia Kristeva, *The Kristeva Reader*, ed. Toril Moi (New York: Columbia University Press, 1986), 94–97, 103–5, 120–22. On the uncanny's relation to the sublime, see Harold Bloom, *Agon: Towards a Theory of Revisionism* (New York: Oxford University Press, 1982), 101–2.
56. Michael Marder, *Plant-Thinking: A Philosophy of Vegetal Life* (New York: Columbia University Press, 2013), 8; Masahiro Mori, "The Uncanny Valley," trans. Karl F. MacDorman and Norri Kageki, IEEE Spectrum, https://spectrum.ieee.org/automaton/robotics/humanoids/the-uncanny-valley, updated June 12, 2012.
57. "Animal Intelligence," *Popular Science News* 24, no. 11 (November 1890): 172; Jean-Paul Sartre, "The Root of the Chestnut Tree," *Partisan Review* 13, no. 1 (Winter 1946): 26, 33.

58. Alasdair MacIntyre, "Existentialism," in *The Encyclopedia of Philosophy*, vol. 3, ed. Paul Edwards (New York: Macmillan and the Free Press, 1967), 149; *The Secret Life of Plants*, dir. Green. See also Marder, *Plant-Thinking*, 37–38.
59. Marder, *Plant-Thinking*, 31–32; Marder and Irigaray, *Through Vegetal Being*, 117; Tompkins and Bird, *Secret Life of Plants*, ix–x.
60. *Secret Life of Plants*, dir. Green.
61. Marder and Irigaray, *Through Vegetal Being*, 158.
62. T. C. N. Singh and A. Gnanam, "Studies on the Effect of Sound Waves of Nashewaram on the Growth and Yield of Paddy," *Journal of Annamalai University* 26B (1965): 78–81; Tompkins and Bird, *Secret Life of Plants*, 146–52, 158. See also Margaret Collins and John E. K. Foreman, "The Effect of Sound on the Growth of Plants," *Canadian Acoustics* 29, no. 2 (2001): 4.
63. Dorothy Retallack, *The Sound of Music and Plants* (Santa Monica, CA: DeVorss, 1973), 17, 19–21, 23, 80. The rock music consisted of the full albums of *Led Zeppelin II* and Jimi Hendrix's *Band of Gypsys*, plus "Break Song" by Vanilla Fudge.
64. T. Olga Curtis, "Music That Kills Plants," *Denver Post Empire Magazine*, June 21, 1970, 8; Retallack, *The Sound of Music and Plants*, 44, 33; Tompkins and Bird, *Secret Lives of Plants*, 156–57; "*CBS Evening News* for 1970-10-16," Vanderbilt Television News Archive, https://tvnews.vanderbilt.edu/programs/206441. In her book, Retallack reported receiving a letter "from two boys in a church-related college who heartily disagreed with the implications of the earlier C.B.S. report. They told me that they had three marijuana plants growing on top of a stereo speaker which plays nothing but rock 24 hours a day and the plants were thriving beautifully." Retallack, *The Sound of Music and Plants*, 42.
65. Tompkins and Bird, *Secret Lives of Plants*, 154–57; Retallack, *The Sound of Music and Plants*, 24, 27, 29.
66. Retallack, *The Sound of Music and Plants*, 59, 65, 68, 75; McKenna, *True Hallucinations*, 84.
67. Retallack, *The Sound of Music and Plants*, 45, 75.

CHAPTER FIVE

1. Terence McKenna, *True Hallucinations: Being an Account of the Author's Extraordinary Adventures in the Devil's Paradise* (San Francisco: HarperSanFrancisco, 1993), 68, 150.
2. Pauline Oliveros, *Sonic Meditations* (Baltimore: Smith Publications, 1974), https://monoskop.org/images/0/09/Oliveros_Pauline_Sonic_Meditations_1974.pdf.
3. Pauline Oliveros, *Software for People: Collected Writings, 1963–1980* (Baltimore: Smith Publications, 1984), 105; Oliveros, *Sonic Meditations*; Pauline Oliveros, interview by Robert Ashley in *Music with Roots in the Aether: Pauline Oliveros*, dir. Robert Ashley (New York: Lovely Music, 1987).

4. Oliveros, *Software for People*, 104.
5. Mitchell Morris, "Sight, Sound, and the Temporality of Myth Making in *Koyaanisqatsi*," in *Beyond the Soundtrack: Representing Music in Cinema*, ed. Daniel Goldmark, Lawrence Kramer, and Richard Leppert (Berkeley: University of California Press, 2007), 125–26.
6. Joan La Barbara, "Philip Glass and Steve Reich: Two from the Steady State School," in *Writings on Glass: Essays, Interviews, Criticism*, ed. Richard Kostelanetz and Robert Flemming (New York: Schirmer Books, 1997), 39; Tom Johnson, "Philip Glass's New Parts," *Village Voice*, April 6, 1972, http://www.villagevoice.com/music/clip-job-tom-johnsons-original-1972-voice-review-of-philip-glasss-music-in-12-parts-6597183.
7. Michael Nyman, *Experimental Music: Cage and Beyond*, 2nd ed. (New York: Cambridge University Press, 1999), n.p., 6–8.
8. Roger Johnson, ed., *Scores: An Anthology of New Music* (New York: Schirmer, 1981), xiv–xv. In his preface to the second edition of *Experimental Music*, Nyman agreed with this assessment, stating that experimental music started in the United States and spread to England, but was distinct from the continental tradition (Nyman, *Experimental Music*, xvii). See also Brian Eno, "Foreword," in Nyman, *Experimental Music*, xi. For a more complete summary of the musicological discourse differentiating the "experimental" music of the sixties and seventies from the modernist avant-garde organized around serialism, see Georgina Born, *Rationalizing Culture: IRCAM, Boulez, and the Institutionalization of the Musical Avant-Garde* (Berkeley: University of California Press, 1995), 48–64.
9. R. Murray Schafer, *The Book of Noise* (Vancouver: Price Print, 1970), 3.
10. R. Murray Schafer, *The New Soundscape: A Handbook for the Modern Music Teacher* (Don Mills, Ontario: BMI Canada, 1969), 58–61.
11. J. Michael McCloskey, *In the Thick of It: My Life in the Sierra Club* (Washington, DC: Island Press, 2005), 113; James T. Bennett, *Corporate Welfare: Crony Capitalism That Enriches the Rich* (New Brunswick, NJ: Transaction, 2015), 68; Joel Primack and Frank Von Hippel, "Scientists, Politics, and SST: A Critical Review," *Bulletin of the Atomic Scientists* 28, no. 4 (April 1972): 29.
12. Quoted in Bennett, *Corporate Welfare*, 57–58, 60.
13. Vine Deloria Jr., *The Metaphysics of Modern Existence* (New York: Harper & Row, 1979), 160.
14. U.S. Environmental Protection Agency, "Hear Here" (Washington, DC: EPA, 1980), 5.
15. R. Murray Schafer, *The Tuning of the World* (New York: Knopf, 1977), 246; Schafer, *Book of Noise*, 14–15, 22.
16. Schafer, *Tuning of the World*, 4, 208; Schafer, *New Soundscape*, 10–11.
17. R. Murray Schafer, "Music of the Environment," in *Audio Culture: Readings in Modern Music*, ed. Christoph Cox and Daniel Warner (New York: Continuum, 2004), 32–33; Schafer, *New Soundscape*, 46; Schafer, *Book of Noise*, 27; Schafer, *Tuning of the World*, 272.

18. Schafer, *Tuning of the World*, 4; Schafer, "Music of the Environment," 30.
19. Schafer, *Book of Noise*, 3; Schafer, *New Soundscape*, 6–8; Schafer, *Tuning of the World*, 216; Brandon LaBelle, *Background Noise: Perspectives on Sound Art* (New York: Continuum, 2006), 210–11.
20. Schafer, *Tuning of the World*, 235–36. An analysis of the recordings was published in 1977 as *Five Village Soundscapes*; the book's cover, depicting a medieval church amidst fields and mountains, attests to the World Soundscape Project's preservationist, bucolic, anti-urban, and anti-industrial leanings. See R. Murray Schafer, *Five Village Soundscapes* (Vancouver: A.R.C. Publications, 1977).

    These ideas also found currency in the work of some American composers. Annea Lockwood began recording the soundscapes of rivers in the late 1960s. Lockwood endeavored to create a River Archive, and beginning in 1974 used her recordings to produce sound installations titled *Play the Ganges Backwards One More Time, Sam*. In 1982 the Hudson River Museum commissioned her to create *A Sound Map of the Hudson River.*
21. Gregory Bateson, *Steps to an Ecology of Mind: Collected Essays in Anthropology, Psychiatry, Evolution, and Epistemology* (Chicago: University of Chicago Press, 1972), 159–60.
22. Pauline Oliveros, in Walter Zimmermann, *Desert Plants: Conversations with 23 American Musicians* (Vancouver: W. Zimmermann: A.R.C. Publications, 1976), n.p.
23. Oliveros, *Software for People*, 28, 85–86; R. Murray Schafer, *Ear Cleaning: Notes for an Experimental Music Course* (Toronto: Berandol Music Limited, 1967), 10, 14.
24. Oliveros, *Software for People*, 85–86. Pychoacoustically speaking, sound level, loudness, and noise level are three different phenomena; and different scales, each with its own units, have been developed to measure them. See Theodore John Schultz, *Community Noise Ratings* (London: Applied Science Publishers, 1972), 2, 6, 12–13, 15.
25. Max Neuhaus, "BANG, BOOooom, ThumP, EEEK, tinkle," *New York Times*, December 6, 1974, http://www.nytimes.com/1974/12/06/archives/bang-booooom-thump-eeek-tinkle.html?_r=0 (emphases mine).
26. Max Neuhaus, "Sound Design," http://www.max-neuhaus.info/soundworks/vectors/invention/sounddesign/; Liz Kotz, "Max Neuhaus: Sound into Space," in *Max Neuhaus*, ed. Max Neuhaus and Christoph Cox (New York: Dia Center for the Arts, 2010), 104; Max Neuhaus, "Listen," http://www.max-neuhaus.info/soundworks/vectors/walks/LISTEN/LISTEN2.htm.
27. Arne Næss, "The Shallow and the Deep, Long-Range Ecology Movement: A Summary," *Inquiry* 16, no. 1 (February 1973): 95.
28. Alvin Lucier and Douglas Simon, *Chambers: Scores by Alvin Lucier* (Middletown, CT: Wesleyan University Press, 2012), 143–44.
29. Lucier and Simon, *Chambers*, 111–16, 118, 121–23.

30. Richard Kostelanetz, "Inferential Art," in *John Cage*, ed. R. Kostelanetz (New York: Praeger, 1970), 107–8; John Cage, *Silence: Lectures and Writings* (London: Marion Boyars, 1995), 4, 8, 12; Nyman, *Experimental Music*, 26.
31. Nyman, *Experimental Music*, 25; Cage, *Silence*, 10.
32. *Walkthrough* was installed at the Jay Street/Borough Hall subway entrance. Thomas Willis, "'Public Supply': A Sound Idea," *Chicago Tribune*, May 2, 1973, quoted in Tracy Fitzpatrick, *Art and the Subway: New York Underground* (New Brunswick, NJ: Rutgers University Press, 2009), 227; Max Neuhaus, "Place and Moment 2," http://www.max-neuhaus.info/soundworks/vectors/place/notes/notes2.htm; Max Neuhaus, "Evocare," http://www.max-neuhaus.info/soundworks/vectors/place/evocare, 2.
33. Zimmermann, *Desert Plants*, n.p.; Tom Johnson, "Richard Teitelbaum on the Threshold," *Village Voice*, June 2, 1975, http://tvonm.editions75.com/articles/1975/richard-teitelbaum-on-the-threshold.html.
34. Pauline Oliveros, "Animals Are Deep Listeners," in *Deep Listening: A Composer's Sound Practice* (New York: Universe, 2005), xxv; Ashley, *Music with Roots in the Aether: Pauline Oliveros*; Johnson, *Scores*, 4; Oliveros, *Software for People*, 30.
35. Oliveros, in Zimmermann, *Desert Plants*, n.p.; Julia Kristeva, *The Kristeva Reader*, ed. Toril Moi (New York: Columbia University Press, 1986), 94–99; La Monte Young, "Dream Music," *Aspen* 9 (Winter–Spring 1967): n.p.
36. Riley was also working with Oliveros in an improvisational ensemble, having been a classmate of hers a couple of years earlier at San Francisco State College. Jeremy Grimshaw, *Draw a Straight Line and Follow It: The Music and Mysticism of La Monte Young* (New York: Oxford University Press, 2011), 88; Keith Potter, *Four Musical Minimalists: La Monte Young, Terry Riley, Steve Reich, Philip Glass* (New York: Cambridge University Press, 2000), 42, 95, 104; William Duckworth, *Talking Music: Conversations with John Cage, Philip Glass, Laurie Anderson, and Five Generations of American Experimental Composers* (New York: Schirmer, 1995), 269–70, 273; Terry Riley, "Obsessed and Passionate about All Music," interviewed by Frank J. Oteri, February 16, 2001, Wortham Theater Center, Houston, TX, 4, https://2104310a1da50059d9c5-d1823d6f516b5299e7df5375e9cf45d2.ssl.cf2.rackcdn.com/nmbx/assets/26/riley_interview.pdf; David Bernstein, ed., *The San Francisco Tape Music Center: 1960s Counterculture and the Avant-Garde* (Berkeley: University of California Press, 2008), 10–11; La Monte Young, "Lecture, 1960," in *Happenings and Other Acts*, ed. Mariellen Sanford (New York: Routledge, 1995), 65 (emphasis mine).
37. Owen Jorgensen, *Tuning: Containing the Perfection of Eighteenth-Century Temperament, the Lost Art of Nineteenth-Century Temperament, and the Science of Equal Temperament* (East Lansing: Michigan State University Press, 1991).
38. Grimshaw, *Draw a Straight Line and Follow It*, 169; Potter, *Four Musical Minimalists*, 116; Riley, "Obsessed and Passionate," 8.

39. Bernstein, *San Francisco Tape Music Center*, viii; Grimshaw, *Draw a Straight Line and Follow It*, 25; Peter Lavezzoli, *The Dawn of Indian Music in the West* (New York: Continuum, 2006), 238. The tambura is a stringed instrument that supplies a continuous drone by cycling four strings in a repeated loop, producing overtone harmonics that enhance the melody of the raga through stochastic resonance.
40. Gary Snyder, who studied Zen Buddhism as a disciple for nine years in Japan, explained the virtues of the master-disciple relationship: "One of the reasons that you have to be very patient and very committed is that the way the transmission works is that you don't *see* how it works for a very long time. It begins to come clear later." Gary Snyder, *The Real Work: Interviews and Talks, 1964–1979* (New York: New Directions, 1980), 98. Grimshaw, *Draw a Straight Line and Follow It*, 105, 118; *In Between the Notes: A Portrait of Pandit Pran Nath*, dir. William Farley (Other Minds Studio, 1986), https://www.youtube.com/watch?v=AYtHhS_Re8Y; Potter, *Four Minimalists*, 136–37. See also https://www.youtube.com/watch?v=nlqXyjwJwYM.
41. Hazrat Inayat Khan, *The Mysticism of Sound and Music: The Sufi Teaching of Hazrat Inayat Khan* (Boston: Shambhala, 1991), 153; Lavezzoli, *Dawn of Indian Music in the West*, 246; Terry Riley, interview by Robert Ashley in *Music with Roots in the Aether: Terry Riley*, dir. Robert Ashley (New York: Lovely Music, 1987). See also Potter, *Four Musical Minimalists*, 103.
42. Martin Clayton, *Time in Indian Music: Rhythm, Metre, and Form in North Indian Rāg Performance* (New York: Oxford University Press, 2008), 13–14; Lavezzoli, *Dawn of Indian Music in the West*, 18; Chetan Karnani, *Listening to Hindustani Music* (Bombay: Orient Longman, 1976), 14.
43. Clayton, *Time in Indian Music*, 10–11, 13–16, 20, 23, 26; George E. Ruckert, *Music in North India: Experiencing Music, Expressing Culture* (New York: Oxford University Press, 2004), 56.
44. Marcus Boon, "Jon Hassell: There Was No Avant Garde," January 1, 2002, http://marcusboon.com/jon-hassell-there-was-no-avant-garde/; Charlemagne Palestine, interview by Brian Duguid, March/April 1996, http://media.hyperreal.org/zines/est/intervs/palestin.html, 1; Andy Battaglia, "Charlemagne Palestine: Strumming Music for Piano, Harpsichord and String Ensemble: Album Review," *Pitchfork*, November 30, 2010, http://pitchfork.com/reviews/albums/14902-strumming-music-for-piano-harpsichord-and-string-ensemble/; Ingram Marshall, liner notes for *Strumming Music* (2010), Forced Exposure, http://www.forcedexposure.com/Catalog/palestine-charlemagne-strumming-music-3cd/SR.297CD.html.
45. Lucier and Simon, *Chambers*, 170; Zimmermann, *Desert Plants*, n.p.; Alvin Lucier, *I Am Sitting in a Room* (Lovely Music VR 1013, 1981).
46. Mike Zwerin, "Max Neuhaus's 'Sound Works' Listen to Surroundings," *New York Times*, November 17, 1999, http://www.nytimes.com/1999/11/

17/style/17iht-max.t.html; Max Neuhaus, in *Max Neuhaus—Times Square*, dir. Rory Logsdail (New York: Firefly Pictures, 2002), http://www.max-neuhaus.info/timessquare.htm; Raymond Ericson, "Max Neuhaus Music Made Especially for Times Square," *New York Times*, November 27, 1977, http://www.nytimes.com/1977/11/27/archives/max-neuhaus-music-made-especially-for-times-square.html?_r=1. The work is still active as of this writing, between Seventh Avenue and Broadway and Forty-Fifth and Forty-Sixth Streets.

47. Max Neuhaus, liner notes for *Max Neuhaus, Fontana Mix—Feed (Six Realizations of John Cage)* (Milan, Italy: Alga Marghen, 2003) (emphasis mine); David Ernst, *The Evolution of Electronic Music* (New York: Macmillan, 1977), 161–62.
48. Nicolas Collins, "'Pea Soup'—a History" (unpublished manuscript), 4, 6, http://www.nicolascollins.com/texts/peasouphistory.pdf. The audio-recording of a 2004 realization is available at http://www.nicolascollins.com/music/peasoupapestaartje.mp3.
49. Nyman, *Experimental Music*, 6.
50. Potter, *Four Musical Minimalists*, 113–15; Alfred Frankenstein, "Music Like None Other on Earth," *San Francisco Chronicle*, November 8, 1964, 28, quoted in Potter, *Four Musical Minimalists*, 149.
51. Oliveros, *Software for People*, 72; Riley, "Obsessed and Passionate," 6 (emphasis mine).
52. Brian Eno, "Generating and Organizing Variety in the Arts," in *Audio Culture*, ed. Cox and Warner, 228, 230; Nyman, *Experimental Music*, 125. Another example of this compositional technique is Michael Nyman's *1–100*, released on Brian Eno's Obscure label in 1976: four pianists play the same sequence of one hundred chords, but the timing is not predetermined. A player moves on to the next chord only when she can no longer hear her last one; this is an individual judgment dependent on both psychophysiological and environmental factors.
53. Eno, "Generating and Organizing Variety in the Arts," 228–30.
54. Brian Eno and Russell Mills, *More Dark than Shark* (London: Faber and Faber, 1986), 74–75; Brian Eno, *Imaginary Landscapes* (New York: Mystic Fire Video, 1989).
55. Potter, *Four Musical Minimalists*, 134; Richard Kostelanetz, "Philip Glass," in *Writings on Glass*, ed. Kostelanetz and Flemming, 111; Allan Kozinn, "The Touring Composer as Keyboardist," in ibid., 103; Charles Merrell Berg, "Philip Glass on Composing for Film and Other Forms: The Case of *Koyaanisqatsi*," in ibid., 131–33; Philip Glass, *Music by Philip Glass* (New York: Harper & Row, 1987), 58–59.
56. David Howard, *Sonic Alchemy: Visionary Music Producers and Their Maverick Recordings* (Milwaukee: Hal Leonard Corporation, 2004), 186; David Sheppard, *On Some Faraway Beach: The Life and Times of Brian Eno* (Chicago: Chicago Review Press, 2009), 38, 50–51, 304; Eno and Mills, *More Dark*

than Shark, 44, 73; Eric Tamm, *Brian Eno: His Music and the Vertical Color of Sound* (Boston: Faber and Faber, 1989), 84; Brian Eno, *A Year with Swollen Appendices: Brian Eno's Diary* (London: Faber and Faber, 1996), 308–9; Brian Eno, liner notes for *Discreet Music* (EG Records Ltd., 1975), http://music.hyperreal.org/artists/brian_eno/discreet-txt.html.

57. Eno, *Imaginary Landscapes*; Eno, *Discreet Music*, liner notes; Eno, "Generating and Organizing Variety in the Arts," 231.
58. Brian Eno, "Ambient Music," in *Audio Culture*, ed. Cox and Warner, 96–97.
59. Eno, "Generating and Organizing Variety in the Arts," 231; Eno and Mills, *More Dark than Shark*, 76; Eno, *Imaginary Landscapes*; Eno, "Ambient Music," 96.

CHAPTER SIX

1. Gregory Bateson, *Steps to an Ecology of Mind: Collected Essays in Anthropology, Psychiatry, Evolution, and Epistemology* (Chicago: University of Chicago Press, 1972), 320; Ervin Laszlo, *Introduction to Systems Philosophy: Toward a New Paradigm of Contemporary Thought* (New York: Gordon and Breach, 1972), 221.
2. Michael W. Fox, *Between Animal and Man* (New York: Coward, McCann & Geoghegan, 1976), 31; Michael W. Fox, "Man and Nature: Biological Perspectives," in *On the Fifth Day: Animal Rights and Human Ethics*, ed. Michael Fox and Richard Morris (Washington, DC: Acropolis Books, 1978), 122–23; Joan McIntyre, *Mind in the Waters: A Book to Celebrate the Consciousness of Whales and Dolphins* (San Francisco: Sierra Club Books, 1974), 8; Peter Singer, *Animal Liberation: A New Ethics for Our Treatment of Animals* (New York: New York Review of Books, 1975), 12, 15.
3. Jane Goodall, *In the Shadow of Man* (Boston: Houghton Mifflin, 1971), 78, 242–49, 267–68; Monks of New Skete, *How to Be Your Dog's Best Friend: A Training Manual for Dog Owners* (Boston: Little, Brown, 1978), xii, 73; Janine Marchessault, *Ecstatic Worlds: Media, Utopias, Ecologies* (Cambridge, MA: MIT Press, 2017), 222, 225; Temple Grandin, *Animals in Translation: Using the Mysteries of Autism to Decode Animal Behavior* (New York: Scribner, 2005), 23.
4. Gary Snyder, *The Real Work: Interviews and Talks, 1964–1979* (New York: New Directions, 1980), 3–4. See also Paul Shepard, *Thinking Animals: Animals and the Development of Human Intelligence* (New York: Viking, 1978), 97.
5. Goodall, *In the Shadow of Man*, 237; Grandin, *Animals in Translation*, 4–5.
6. Cynthia Novack, *Sharing the Dance: Contact Improvisation and American Culture* (Madison: University of Wisconsin Press, 1990), 53, 11; Daniel Lepkoff, "Contact Improvisation, or What Happens When I Focus My Attention on the Sensation of Gravity, the Earth, and My Partner?" *Nouvelles de danse N° 38–39: Contact Improvisation* (March 1999), http://daniellepkoff

.com/writings/CI%20What%20happens%20when.php. In the twenty-first century, other dancers extended contact improvisation to include nonhuman animals—specifically, dolphins and horses. See Nita Little, "Enminded Performance: Dancing with a Horse," in *Sentient Performativities of Embodiment: Thinking Alongside the Human*, ed. Lynette Hunter, Elisabeth Krimmer, and Peter Lichtenfels (Lanham, MD: Lexington Books, 2016), 103; Laura Cull, "From *Homo Performans* to Interspecies Collaboration," in *Performing Animality: Animals in Performance Practices*, ed. Jennifer Parker-Starbuck and Lourdes Orozco García (New York: Palgrave Macmillan, 2015), 30–32; and Una Chaudhuri, *Stage Lives of Animals: Zooesis and Performance* (New York: Routledge, 2017), 196–97, 201.

7. Rex Weyler, *Greenpeace: How a Group of Journalists, Ecologists, and Visionaries Changed the World* (Emmaus, PA: Rodale, 2004), 259; D. Graham Burnett, *The Sounding of the Whale: Science and Cetaceans in the Twentieth Century* (Chicago: University of Chicago Press, 2012), 626; Farley Mowat, *A Whale for the Killing* (Boston: Little, Brown, 1972).
8. Mowat, *A Whale for the Killing*, 189.
9. Masahiro Mori, "The Uncanny Valley," trans. Karl F. MacDorman and Norri Kageki, IEEE Spectrum, https://spectrum.ieee.org/automaton/robotics/humanoids/the-uncanny-valley, updated June 12, 2012; Mowat, *A Whale for the Killing*, 59.
10. Shepard, *Thinking Animals*, 106.
11. Maurice Temerlin, *Lucy, Growing Up Human: A Chimpanzee Daughter in a Psychotherapist's Family* (Palo Alto, CA: Science and Behavior Books, 1975). Koko the gorilla was also taught American Sign Language beginning in 1972 by Francine Patterson of Stanford University, inspired by the Gardners' work with Washoe; see Francine Patterson and Eugene Linden, *The Education of Koko* (New York: Holt, Rinehart and Winston, 1981), http://www.koko.org/sites/default/files/root/pdfs/teok_book.pdf; Eugene Linden, *Apes, Men, and Language* (New York: Saturday Review Press, 1974); Eugene Linden, *Silent Partners: The Legacy of the Ape Language Experiments* (New York: Times Books, 1986), ix, 208–9; Dale Peterson and Jane Goodall, *Visions of Caliban: On Chimpanzees and People* (Athens: University of Georgia Press, 2000), 311.
12. David Hamburg, "Foreword," in Goodall, *In the Shadow of Man*, xvii. Goodall had begun her study in 1960, after Louis Leakey encouraged her to think that the behavior of chimps would shed light on that of Stone Age humans (6).
13. Terrace relied on a definition of language advanced by Noam Chomsky, which was why the chimp was named Neam Chimpsky, or "Nim" for short. Cognitive ethologist Louis Herman, who used a broader criterion of "language learning" in his study of dolphins, commented wryly, "It seems likely . . . that linguists will continually be able to retreat to new definitions of language in the face of demonstrations that animals have met

prior criteria." Louis Herman, "Cognitive Characteristics of Dolphins," in *Cetacean Behavior: Mechanisms and Functions*, ed. Louis Herman (New York: John Wiley & Sons, 1980), 410; Linden, *Silent Partners*, 68–69.

14. W. F. Brewer, "There Is No Convincing Evidence for Operant or Classical Conditioning in Adult Humans," in *Cognition and the Symbolic Processes*, ed. W. B. Weimer and D. S. Palermo (Hillsdale, NJ: Lawrence Erlbaum Associates, 1974), 29, quoted in Donald R. Griffin, *The Question of Animal Awareness: Evolutionary Continuity of Mental Experience* (New York: Rockefeller University Press, 1976), 49; Singer, *Animal Liberation*, 12–13.
15. Karen Pryor, *Lads Before the Wind: Adventures in Porpoise Training* (New York: Harper & Row, 1975), 148; Jacques-Yves Cousteau and Philippe Diolé, *Dolphins*, trans. J. F. Bernard (Garden City, NY: Doubleday, 1975), 158–59; Goodall, *In the Shadow of Man*, 256.
16. John Lilly, "Learning Motivated by Subcortical Simulation: The 'Start' and 'Stop' Patterns of Behavior," in *Reticular Formation of the Brain*, ed. Herbert Jasper et al. (Boston: Little, Brown, 1958), 705; D. Graham Burnett, "Adult Swim," in *Groovy Science: Knowledge, Innovation, and American Counterculture*, ed. David Kaiser and W. Patrick McCray (Chicago: University of Chicago Press, 2016), 15, 25, 37; Burnett, *Sounding of the Whale*, 615; John Lilly, *The Center of the Cyclone: An Autobiography of Inner Space* (New York: Bantam, 1972), 48, 50.
17. John Lilly, *The Mind of the Dolphin: A Nonhuman Intelligence* (Garden City, NY: Doubleday, 1967), 94, xvi, 32.
18. The taxonomic clad of Cetacea includes both the "great whales," such as the humpback, blue, and sperm whales, and smaller species. The cetaceans trained to interact with humans in captivity have been exclusively of the family *Delphinidae* of the suborder *Odontoceti* (toothed whales). These include the genera *Steno* (the rough-toothed dolphin), *Tursiops* (bottlenose dolphin), *Stenella* (spinner, spotted, and striped dolphins), *Delphinus* (common dolphin), *Orcinus* (killer whale), and *Phocoena* (harbor porpoise). Trainers typically called all of them, except perhaps the killer whale, "porpoises." Pryor, *Lads Before the Wind*, 6. All *Delphinidae* caught in tuna fishermen's nets (typically *Stenella*) were also colloquially referred to as "porpoises." Richard Ellis, *Dolphins and Porpoises* (New York: Knopf, 1982), 3, 5.
19. Bateson, *Steps*, 368; Peter S. Fiebleman, "Mike Nichols Tries to Make a 'Talkie' with Dolphins," *Atlantic*, January 1974, 74–76; Ray Hunt, *Think Harmony with Horses: An In-Depth Study of Horse/Man Relationship* (Bruneau, ID: Give-It-A-Go Books, 1978), 38. In publications, dolphin trainers related many instances of autonomous acts of intelligence by dolphins, including teasing, conspiring, playing and inventing games, problem solving, and inventing and using tools. See Antony Alpers, *Dolphins: The Myth and the Mammal* (Boston: Houghton Mifflin, 1961), 91, 93–94, 97, 103, 105; Lilly, *Mind of the Dolphin*, 64; Ellis, *Dolphins and Porpoises*, 65, 75; and Pryor, *Lads Before the Wind*, 215–16.

20. Hunt, *Think Harmony with Horses*, 7; Karen Pryor, "Why Porpoise Trainers Are Not Dolphin Lovers: Real and False Communication in the Operant Setting," *Annals of the New York Academy of Sciences* 364, no. 1 (1981): 138–39; Pryor, *Lads Before the Wind*, 23; Hunt, *Think Harmony with Horses*, 15–16.
21. Hunt, *Think Harmony with Horses*, 5, 7; Karen Pryor, "Behavior and Learning in Porpoises and Whales," *Naturwissenschaften* 60 (1973): 417, 419.
22. Kenneth S. Norris, *The Porpoise Watcher* (New York: Norton, 1974), 155; Hunt, *Think Harmony with Horses*, 2, 5–6, 26, 39.
23. Hunt, *Think Harmony with Horses*, 38; Pryor, *Lads Before the Wind*, 124–27; Pryor, "Why Porpoise Trainers," 141.
24. Gregory Bateson, "Observation of a Cetacean Community," in McIntyre, *Mind in the Waters*, 142; Burnett, "Adult Swim," 34; Pryor, *Lads Before the Wind*, 155, 171–72, 236–42, 252–53; Karen Pryor, Richard Haag, and Joseph O'Reilly, "The Creative Porpoise: Training for Novel Behavior," *Journal of the Experimental Analysis of Behavior* 12, no. 4 (July 1969): 654–61; Stewart Brand, "Both Sides of the Necessary Paradox," *Harper's*, November 1973, 34; Pryor, "Behavior and Learning in Porpoises," 417.
25. Pryor, *Lads Before the Wind*, 124, 127, 218; Konrad Lorenz, in ibid., ix.
26. Hunt, *Think Harmony with Horses*, 2–3; Robert Hunter, *Warriors of the Rainbow: A Chronicle of the Greenpeace Movement from 1971 to 1979* (Fremantle, WA: Greenpeace in association with Fremantle Press, 2011 [1979]), 138–41, 228–39; Weyler, *Greenpeace*, 207–10.
27. Norris, *Porpoise Watcher*, 247–49.
28. These books include Victor Scheffer's *The Year of the Whale*, Karen Pryor's *Lads Before the Wind*, Ken Norris's *The Porpoise Watcher*, Joan McIntyre's *Mind in the Waters*, John Lilly's *Lilly on Dolphins*, Robin Brown's *The Lure of the Dolphin*, Jacques Cousteau's *Dolphins*, and Richard Ellis's *Dolphins and Porpoises*. On dolphins and telepathy, see Robin Brown, *The Lure of the Dolphin* (New York: Avon, 1979), 38–42.
29. Lilly, *Mind of the Dolphin*, 134.
30. Weyler, *Greenpeace*, 207–9; Burnett, *Sounding of the Whale*, 529; Hunter, *Warriors of the Rainbow*, 139; *Greenpeace: Voyages to Save the Whales* (Canada: Greenpeace Film Productions, 1977).
31. Roger Payne and Scott McVay, "Songs of Humpback Whales," *Science* 173, no. 3997 (August 13, 1971): 585–97; Roger Payne, "Humpbacks: Their Mysterious Songs," *National Geographic*, January 1979, 19; Bateson, *Steps*, 375.
32. Burnett, *Sounding of the Whale*, 524–26, 630, 633, 654; Gary Snyder, *The Old Ways: Six Essays* (San Francisco: City Lights Books, 1977), 13. In November 1970, the U.S. secretary of interior (Walter Hickel) placed all the large whales on the Endangered Species List, making the importation of any whale product into the United States a crime. The international whaling moratorium was finally negotiated in 1982, with Japan, Peru, Norway, and the Soviet Union dissenting.

33. Alvin Lucier and Douglas Simon, *Chambers: Scores by Alvin Lucier* (Middletown, CT: Wesleyan University Press, 2012), 64.
34. Payne, "Humpbacks," 18.
35. Hunter, *Warriors of the Rainbow*, 141–42.
36. Maurice Burton, *The Sixth Sense of Animals* (New York: Taplinger, 1972), 2–6, 174; Norris, *Porpoise Watcher*, 209. See also Gilles Deleuze, *The Logic of Sense*, trans. Mark Lester with Charles Stivale, ed. Constantin Boundas (New York: Columbia University Press, 1990).
37. Richard O'Barry, *Behind the Dolphin Smile* (Chapel Hill, NC: Algonquin Books, 1988), 102–3.
38. Barbara B. Brown, *New Mind, New Body: Bio-Feedback: New Directions for the Mind* (New York: Harper & Row, 1974), 74–75, 79; Elizabeth Dodson Gray, *Green Paradise Lost* (Wellesley, MA: Roundtable Press, 1979), 93.
39. Snyder, *Real Work*, 107; Shepard, *Thinking Animals*, 97. In this connection, choreography is broadly defined as any self-aware movement through space. See David Gere, *How to Make Dances in an Epidemic: Tracking Choreography in the Age of AIDS* (Madison: University of Wisconsin Press, 2004), 9.
40. Lucier and Simon, *Chambers*, 16–17, 22.
41. Burnett, "Adult Swim," 24; Lilly, *Center of the Cyclone*, 103.
42. Lilly, *Center of the Cyclone*, 60–61, 91, 103.
43. McIntyre, *Mind in the Waters*, 217–18, 220.
44. Simone Forti, *Handbook in Motion* (Northampton, MA: Contact Editions, 1974), 29, 68, 91. Rebekah Kowal, *How to Do Things with Dance: Performing Change in Postwar America* (Middletown, CT: Wesleyan University Press, 2010), 233; Julia Bryan-Wilson, "Animate Matters: Simone Forti in Rome," in *Simone Forti: Thinking with the Body*, ed. Sabine Breitweiser (Salzburg: Museum der Moderne, 2014), 56.
45. Forti, *Handbook*, 91; Bryan-Wilson, "Animate Matters," 53–54; Simone Forti, "Dancing at the Fence," *Avalanche* 10 (December 1974): 22–23. Forti had been born in Italy and had fled as a child during World War II to the United States (her family was Jewish).
46. Liz Kotz, "Simone Forti and Sound," in *Simone Forti*, ed. Breitweiser, 64.
47. Forti, *Handbook*, 64, 18–19, 119.
48. Forti, *Handbook*, 17.
49. Forti, *Handbook*, 136, 129; Forti, "Dancing at the Fence," 20; Bryan-Wilson, "Animate Matters," 56; Sally Banes, *Terpsichore in Sneakers: Post-Modern Dance* (Boston: Houghton Mifflin, 1980), 36. See also *Simone Forti*, ed. Breitweiser, 10, 21, 26, 30, 32–33, 200.
50. Novack, *Sharing the Dance*, 3, 12.
51. Novack, *Sharing the Dance*, 152–53, 183.
52. Max Neuhaus, liner notes for *Fontana Mix-Feed* (Alga Marghen, 2003); Novack, *Sharing the Dance*, 60; Steve Paxton, "Origins of the Small Dance," interview with Nathan Wagoner, August 23, 2008, https://web.archive.org/web/20121218051805/http://vimeo.com/1731742.

53. Novack, *Sharing the Dance*, 70–71, 152–53, 189 (emphasis mine).
54. Hunt, *Think Harmony with Horses*, 67–68.
55. Buck Brannaman and William Reynolds, *Faraway Horses: Adventures and Wisdom of an American Horse Whisperer* (New York: Lyons, 2003), 67; Robert Miller and Rick Lamb, *The Revolution in Horsemanship and What It Means to Mankind* (Guilford, CT: Lyons Press, 2005), 5; Hunt, *Think Harmony with Horses*, 2.
56. Hunt, *Think Harmony with Horses*, 1, 15–16, 18, 27, 43.
57. Hunt, *Think Harmony with Horses*, 13, 23, 25.
58. Robert Redford, "Riding the Outlaw Trail," *National Geographic*, November 1976, 622–57; Robert Redford, *Outlaw Trail: A Journey Through Time* (New York: Grosset & Dunlap, 1978).
59. Brannaman and Reynolds, *Faraway Horses*, 61, 16–17, 122–23; *Buck*, dir. Cindy Meehl, DVD (IFC Films, 2011).
60. Amanda Uechi Ronan, "Celebrity Equestrian: Robert Redford," *Horse Nation*, June 26, 2014, http://www.horsenation.com/2014/06/26/celebrity-equestrian-robert-redford/. In 1998 Redford played Brannaman in the movie *The Horse Whisperer.*
61. Brannaman and Reynolds, *Faraway Horses*, 67. Similarly, revisionist Westerns of the early seventies, like *Little Big Man* (1970) and *McCabe & Mrs. Miller* (1971), recast the "winning of the West" as an unethical exploitation of vulnerable people and nature.
62. Bruce J. Schulman, *The Seventies: The Great Shift in American Culture, Society, and Politics* (New York: Free Press, 2001), 177.
63. Shepard, *Thinking Animals*, 92–94, 161, 165.
64. Goodall, *In the Shadow of Man*, 256; Bateson, *Steps*, 377; John Lilly, *Lilly on Dolphins: Humans of the Sea* (Garden City, NY: Anchor Books, 1975), xiv.
65. Rob Nixon defines slow violence as "a violence that occurs gradually and out of sight, a violence of delayed destruction that is dispersed across time and space, an attritional violence that is typically not viewed as violence at all." Rob Nixon, *Slow Violence and the Environmentalism of the Poor* (Cambridge, MA: Harvard University Press), 2.
66. Gregory Stephens, "*Koyaanisqatsi* and the Visual Narrative of Environmental Film," *Screening the Past* 28 (Fall 2010): 20–21, http://www.screeningthepast.com/2015/01/koyaanisqatsi%C2%A0and-the-visual-narrative-of-environmental-film/. See also Carl Sagan, *The Dragons of Eden: Speculations on the Evolution of Human Intelligence* (New York: Random House, 1977), 120.
67. Michael Sragow, "An Honest Bull Session with Carroll Ballard," Criterion Collection, https://www.criterion.com/current/posts/4671-an-honest-bull-session-with-carroll-ballard.
68. Roderick Nash, *The Rights of Nature: A History of Environmental Ethics* (Madison: University of Wisconsin Press, 1989), 187–89. Native American (Eskimo) whaling was another divisive issue; see Richard Ellis, *The Book of Whales* (New York: Knopf, 1985), 87–89.

69. Nash, *Rights of Nature,* 181, 190–91. In 1980 David Foreman, who had held leadership positions in both the Wilderness Society and the Nature Conservancy, left them in frustration to found Earth First!, an organization dedicated to sabotaging industrial incursions into wilderness areas. Earth First! was inspired by Edward Abbey's 1975 novel, *The Monkey Wrench Gang,* and Abbey contributed a foreword to the organization's manual *Ecodefense: A Field Guide to Monkeywrenching* (Tucson: N. Ludd, 1987). Oates, *Earth Rising,* 193.
70. Cousteau and Diolé, *Dolphins,* 25, 57, 60, 76–77, 123, 158–60. When Ric O'Barry attempted to liberate a dolphin that he had previously captured, he could not convince it to leave its enclosure. See O'Barry, *Behind the Dolphin Smile,* 5–7, 23.
71. Arthur Lubow, "Riot in Fish Tank 11," *New Times,* October 14, 1977, 38–40.
72. Lubow, "Riot in Fish Tank," 40, 42.
73. Lubow, "Riot in Fish Tank," 40.
74. Gavan Daws, "'Animal Liberation' as Crime: The Hawaii Dolphin Case," in *Ethics and Animals,* ed. Harlan Miller and William H. Williams (Clifton, NJ: Humana Press, 1983), 361.
75. Lubow, "Riot in Fish Tank," 42.
76. Lubow, "Riot in Fish Tank," 52.
77. Lubow, "Riot in Fish Tank," 51–52.
78. Daws, "'Animal Liberation,'" 371; "Value of 2 Dolphins Set Free in '77 at Issue in Hawaii Case," *New York Times,* March 8, 1982.

CHAPTER SEVEN

1. Norbert Wiener, *Cybernetics: Or Control and Communication in the Animal or Machine,* 2nd ed. (Cambridge, MA: MIT Press, 1961), 158–59.
2. Richard Dawkins, *Unweaving the Rainbow: Science, Delusion, and the Appetite for Wonder* (Boston: Houghton Mifflin, 1998), 221–24; W. Ford Doolittle, "Is Nature Really Motherly?" *CoEvolution Quarterly,* Spring 1981, 58–63.
3. Lynn Margulis, "Gaia Is a Tough Bitch," in *The Third Culture: Beyond the Scientific Revolution,* ed. John Brockman (New York: Simon & Schuster, 1995), 130, 132–33.
4. Lynn Margulis, "Lynn Margulis Responds," *CoEvolution Quarterly,* Spring 1981, 64; James Lovelock, "James Lovelock Responds," *CoEvolution Quarterly,* Spring 1981, 62; Margulis, "Gaia Is a Tough Bitch," 134.
5. Edward O. Wilson, *Sociobiology: The New Synthesis* (Cambridge, MA: Harvard University Press, 1975), 4, 11, 28, 109; Ullica Segerstråle, *Defenders of the Truth: The Battle for Science in the Sociobiology Debate and Beyond* (New York: Oxford University Press, 2000), 37, 132–33; Edward O. Wilson, *On Human Nature* (Cambridge, MA: Harvard University Press, 1978), 167.
6. Joel B. Hagen, *An Entangled Bank: The Origins of Ecosystem Ecology* (New Brunswick, NJ: Rutgers University Press, 1992), 155–57; William D.

Hamilton, "Innate Social Aptitudes of Man: An Approach from Evolutionary Genetics," in *Biosocial Anthropology*, ed. Robin Fox (London: Malaby Press, 1975), quoted in Segerstråle, *Defenders of the Truth*, 56; Wilson, *Sociobiology*, 110.

7. Segerstråle, *Defenders of the Truth*, 56, 60, 66, 68, 85; John Maynard Smith, *Evolution and the Theory of Games* (New York: Cambridge University Press, 1982), 1–2, 11, 16.
8. Wilson, *Sociobiology*, 7; Wilson, *On Human Nature*, 167.
9. C. H. Waddington, "Mindless Societies," in *The Sociobiology Debate: Readings on Ethical and Scientific Issues*, ed. Arthur L. Caplan (New York: Harper & Row, 1978), 257; Elizabeth Allen et al., "Against Sociobiology," *New York Review of Books*, November 13, 1975, http://www.nybooks.com/articles/1975/11/13/against-sociobiology/; Marshall Sahlins, *The Use and Abuse of Biology: An Anthropological Critique of Sociobiology* (Ann Arbor: University of Michigan Press, 1976), xii, xiv, 9, 60 (emphasis mine).
10. Segerstråle, *Defenders of the Truth*, 23, 142, 144. These critics claimed vindication in 1979, when right-wing groups in both England and France cited sociobiology to justify their views that racism and human hierarchy were natural, inevitable, and therefore advisable as national policy. See David Dickson, "Sociobiology Critics Claim Fears Come True," *Nature* 282 (November 22, 1979): 348.
11. Edward O. Wilson, "Academic Vigilantism," in *The Sociobiology Debate*, ed. Caplan, 291; Richard Dawkins, *The Selfish Gene* (New York: Oxford University Press, 1989), 2–3.
12. Dawkins, *Selfish Gene*, 200–201; Richard Dawkins, "A Survival Machine," in *Third Culture*, ed. Brockman, 78.
13. Stephen Jay Gould, "Is a New and General Theory of Evolution Emerging?" *Paleobiology* 6, no. 1 (Winter 1980): 121, 125–26. See also Gregory Bateson, *Mind and Nature: A Necessary Unity* (New York: Dutton, 1979), 169–72.
14. Bateson, *Mind and Nature*, 5, 28, 92, 126–27.
15. Andrew Brown, *The Darwin Wars: How Stupid Genes Became Selfish Gods* (London: Simon & Schuster, 1999), 25; Segerstråle, *Defenders of the Truth*, 94, 158. By the mid-1980s, Wilson, Dawkins, and Gould had all qualified their positions to the point that their substantive differences were difficult to discern. See Charles Lumsden and Edward O. Wilson, *Genes, Mind, and Culture: The Co-Evolutionary Process* (Cambridge, MA: Harvard University Press, 1981); and Richard Dawkins, *The Extended Phenotype: The Long Reach of the Gene* (New York: Oxford University Press, 1983). See also Brian Goodwin and Peter Saunders, *Theoretical Biology: Epigenetic and Evolutionary Order from Complex Systems* (Edinburgh: Edinburgh University Press, 1989).
16. "Paranormal Phenomena Facing Scientific Study," *New York Times*, May 1, 1976, 26. See also Carl Sagan, *The Demon-Haunted World: Science as a Candle in the Dark* (London: Headline, 1997), 25, 48–49, 102, 211–12, 253.

17. James E. Horigan, *Chance or Design?* (New York: Philosophical Library, 1979), 167–70; Stephen Jay Gould, *Hen's Teeth and Horse's Toes: Further Reflections in Natural History* (New York: Norton, 1983), 260; "Reagan Says He Questions Evolution," *Dallas Times Herald*, August 23, 1980, Metro section, 1.
18. Philip Mirowski, *Machine Dreams: Economics Becomes a Cyborg Science* (New York: Cambridge University Press, 2002), 129, 204; William Poundstone, *Prisoner's Dilemma: John von Neumann, Game Theory, and the Puzzle of the Bomb* (New York: Doubleday, 1992), 231; J. Craig Jenkins and Craig M. Eckert, "The Right Turn in Economic Policy: Business Elites and the New Conservative Economics," *Sociological Forum* 15, no. 2 (June 2000): 308–15; David Harvey, *A Brief History of Neoliberalism* (Cambridge: Oxford University Press, 2005), 14, 23.
19. Smith named this the Hayek hypothesis. Vernon L. Smith, *Papers in Experimental Economics* (New York: Cambridge University Press, 1991), 223, quoted in Mirowski, *Machine Dreams*, 547; William H. Miernyk, *The Illusions of Conventional Economics* (Morgantown: West Virginia University Press, 1982), 24–25; "Economic Recovery Tax Act of 1981," Pub. L. 97-34, 95 Stat. 172, http://www.legisworks.org/GPO/STATUTE-95-Pg172.pdf.
20. William Tucker, *Progress and Privilege: America in the Age of Environmentalism* (Garden City, NY: Anchor, 1982), 223; George Bush, quoted in Paul Sabin, *The Bet: Paul Ehrlich, Julian Simon, and Our Gamble Over Earth's Future* (New Haven, CT: Yale University Press, 2013), 160.
21. By contrast, James Lovelock wrote in 1981 in the pages of *CoEvolution Quarterly*, "There is only one pollution, namely people. When there are too many of them almost anything, even eventually breathing, can be a pollution" (Lovelock, "James Lovelock Responds," 63). Kristin Luker, *Abortion and the Politics of Motherhood* (Berkeley: University of California Press, 1984), 194–98; Tucker, *Progress and Privilege*, 217; Julian Simon, *The Ultimate Resource* (Princeton, NJ: Princeton University Press, 1981), 337, 345; Sabin, *The Bet*, x, 156, 185–86, 188, 201.
22. Sharon Ghamari-Tabrizi, *The Worlds of Herman Kahn: The Intuitive Science of Thermonuclear War* (Cambridge, MA: Harvard University Press, 2005), 309; Julian Simon and Herman Kahn, eds., *Resourceful Earth: A Response to Global 2000* (New York: Blackwell, 1984), 3; Sabin, *The Bet*, 5, 102.
23. Steve Joshua Heims, *The Cybernetics Group* (Cambridge, MA: MIT Press, 1991), 9; Robert Axelrod, *The Evolution of Cooperation* (New York: Basic Books, 1984), 2, 6; Manfred Drack, "On the Making of a System Theory of Life: Paul A. Weiss and Ludwig von Bertalanffy's Conceptual Connection," *Quarterly Review of Biology* 82, no. 4 (December 2007): 359; Sabin, *The Bet*, 103, 112; R. C. Lewontin, Steven Rose, and Leon J. Kamin, *Not in Our Genes: Biology, Ideology, and Human Nature* (New York: Pantheon, 1984), 4.

24. David Harvey describes neoliberalism as a philosophy of "liberating individual entrepreneurial freedoms" within an "institutional framework characterized by strong private property rights" (Harvey, *Brief History of Neoliberalism*, 2–3). Lewontin, Rose, and Kamin, *Not in Our Genes*, 4; Michael Rogin, *"Ronald Reagan," the Movie: And Other Episodes in Political Demonology* (Berkeley: University of California Press, 1987), xvii. See also Philip Jenkins, *Decade of Nightmares: The End of the Sixties and the Making of Eighties America* (New York: Oxford University Press, 2006), 11–12.
25. Thomas Hine, *The Great Funk: Falling Apart and Coming Together (on a Shag Rug) in the Seventies* (New York: Farrar, Straus and Giroux, 2007), 219–20; Bruce J. Schulman, *The Seventies: The Great Shift in American Culture, Society, and Politics* (New York: Free Press, 2001), 247, 227; Paul N. Edwards, *The Closed World: Computers and the Politics of Discourse in Cold War America* (Cambridge, MA: MIT Press, 1996), 288.
26. William Adams, "American Gothic: *Country, The River, Places in the Heart,*" *Antioch Review* 43, no. 2 (Spring 1985): 221.
27. Adams, "American Gothic," 218–19, 222, 224; William Adams, "Natural Virtue: Symbol and Imagination in the American Farm Crisis," *Georgia Review* 39, no. 4 (Winter 1985): 709.

CONCLUSION

1. Clifford Geertz, *The Interpretation of Cultures* (New York: Basic Books, 1973), 448.
2. Hendrick Hertzberg, "Foreword," in Kevin Mattson, *What the Heck Are You Up to, Mr. President?: Jimmy Carter, America's "Malaise," and the Speech That Should Have Changed the Country* (New York: Bloomsbury, 2010), xiii.
3. Jimmy Carter, "Energy and National Goals," in *Jimmy Carter and the Energy Crisis of the 1970s: "The Crisis of Confidence" Speech of July 15, 1979*, ed. Daniel Horowitz (New York: Bedford/St. Martin's, 2005), 111–19. All subsequent quotations from Carter's speech are also from this source.
4. See Sacvan Bercovitch, *The American Jeremiad* (Madison: University of Wisconsin Press, 1978). On trauma and the fragmentation of identity, see Judith Lewis Herman, *Trauma and Recovery: The Aftermath of Violence—from Domestic Abuse to Political Terror* (New York: Basic Books, 1977), 2, 34–35.
5. Horowitz, *Jimmy Carter and the Energy Crisis*, 17–19.
6. See D. H. Wrong, "The Oversocialized Conception of Man in Modern Sociology," *American Sociological Review* 26, no. 2 (April 1961): 183–93; and James S. House, "Social Structure and Personality," in *Social Psychology: Sociological Perspectives*, ed. Morris Rosenberg and Ralph H. Turner (New Brunswick, NJ: Transaction, 1990), 527, 531–34.
7. Christopher Lasch, *The Culture of Narcissism: American Life in an Age of Diminishing Expectations* (New York: Norton, 1979), xvi, xviii, 23–25; Henry

Allen, "Doomsayer of the Me Decade," *Washington Post*, January 24, 1979, https://www.washingtonpost.com/archive/lifestyle/1979/01/24/doomsayer-of-the-me-decade/57f252a0-5a2b-4cdc-a9ff-c432e9bb2767/?noredirect=on&utm_term=.de772465d26e.

8. Christopher Lasch, *Haven in a Heartless World: The Family Besieged* (New York: Basic, 1977).
9. Tom Wolfe, "The 'Me' Decade and the Third Great Awakening," *New York Magazine*, August 23, 1976, http://nymag.com/news/features/45938/.
10. Christopher Lasch, "The Narcissist Society," *New York Review of Books*, September 30, 1976, https://www.nybooks.com/articles/1976/09/30/the-narcissist-society/.
11. Wolfe, "'Me' Decade."
12. Lasch, *Culture of Narcissism*, 5.
13. Lasch, "Narcissist Society."
14. Allen, "Doomsayer."
15. Thomas Hine, *The Great Funk: Falling Apart and Coming Together (on a Shag Rug) in the Seventies* (New York: Farrar, Straus and Giroux, 2007), 3; Andreas Killen, *1973 Nervous Breakdown: Watergate, Warhol, and the Birth of Post-Sixties America* (New York: Bloomsbury, 2006), 2; Philip Jenkins, *Decade of Nightmares: The End of the Sixties and the Making of Eighties America* (New York: Oxford University Press, 2006), 16, 65–67.
16. Killen, *1973*, 8; Bruce J. Schulman, *The Seventies: The Great Shift in American Culture, Society, and Politics* (New York: Free Press, 2001), xv–xvi.
17. Hine, *Great Funk*, 10–11, 25, 98, 183–84, 34, 15, 27.
18. Hine, *Great Funk*, 13, 27; Schulman, *Seventies*, 80; Jenkins, *Decade of Nightmares*, 26.
19. Wolfe, "'Me' Decade"; Schulman, *Seventies*, 80, xiv, 99; Hine, *Great Funk*, 16–18.
20. Hine, *Great Funk*, 29, 61, 63, 66–67, 72, 74, 86; Schulman, *Seventies*, 88–90, 86, 99–100.
21. Jenkins, *Decade of Nightmares*, 24.
22. Jenkins, *Decade of Nightmares*, 4, 5, 6, 8, 25–26; Lasch, quoted in Allen, "Doomsayer"; Schulman, *Seventies*, 11–12, 18–19.
23. Quoted in Schulman, *Seventies*, 16; Hine, *Great Funk*, 184.
24. Killen, *1973*, 6.
25. Schulman, *Seventies*, 9, xvi.
26. Wolfe, "'Me' Decade"; Lasch, "Narcissist Society"; Killen, *1973*, 2; Schulman, *Seventies*, 13, 261, 19.
27. Hine, *Great Funk*, 33, 19, 194, 25–26; Jenkins, *Decade of Nightmares*, 18–19 (emphases mine).
28. Jenkins, *Decade of Nightmares*, 20; Killen, *1973*, 272; Hine, *Great Funk*, 224.
29. Hine, *Great Funk*, 219–20, 222; Jenkins, *Decade of Nightmares*, 20–23, 9, 11, 12, 14, 30.
30. Schulman, *Seventies*, 220, 257, xiv–xv (emphasis mine).

31. In an airport bookstore, I recently came across the following titles: Lyanda Lynn Haupt, *Mozart's Starling* (2017); David George Haskell, *The Songs of Trees* (2017); and Tom Shroder, *Acid Test* (2015).
32. Eve Kosofsky Sedgwick, *Touching Feeling: Affect, Pedagogy, Performativity* (Durham, NC: Duke University Press, 2003), 130, 143.
33. Jean Baudrillard, "The Precession of Simulacra," in *Media and Cultural Studies: Keyworks*, ed. Meenakshi Durham and Douglas Kellner (Malden, MA: Blackwell, 2001), 534.
34. Killen, *1973*, 69, 152, 166–68, 6–7.
35. Michel Foucault, *The History of Sexuality, Volume 1: An Introduction*, trans. Robert Hurley (New York: Vintage, 1978), 3–7, 18–20; Orit Halpern, *Beautiful Data: A History of Vision and Reason since 1945* (Durham, NC: Duke University Press, 2014), 6; Paul N. Edwards, *The Closed World: Computers and the Politics of Discourse in Cold War America* (Cambridge, MA: MIT Press, 1996), 1–2; Philip Mirowski, *Machine Dreams: Economics Becomes a Cyborg Science* (New York: Cambridge University Press, 2002), 2–3, 6.
36. Sedgwick, *Touching Feeling*, 130, 126.
37. As Paul Weiss wrote in 1977, "Nothing in nature *is* negligible; only *we*, as observers and manipulators of nature, judge and *declare* to be 'negligible' given aspects of nature," according to setting and interest. Paul A. Weiss, "The System of Nature and the Nature of Systems," in *Toward a Man-Centered Medical Science*, ed. Karl E. Schaefer, Herbert Hensel, and Ronald Brady (Mt. Kisco, NY: Futura, 1977), 22–23, 46.
38. On the usable past, see Van Wyck Brooks, "On Creating a Usable Past," in *Van Wyck Brooks: The Early Years*, ed. Claire Sprague, rev. ed. (Boston: Northeastern University Press, 1993 [1918]), 220, 225.

Selected Bibliography

Adams, William. "American Gothic: *Country, The River, Places in the Heart.*" *Antioch Review* 43, no. 2 (Spring 1985): 217–24.

Adler, Margot. *Drawing Down the Moon: Witches, Druids, Goddess-Worshippers, and Other Pagans in America Today.* Rev. and exp. ed. Boston: Beacon Press, 1986.

Akwesasne Notes, ed. *A Basic Call to Consciousness.* Rooseveltown, NY: Mohawk Nation, 1978.

Allen, Henry. "Doomsayer of the Me Decade." *Washington Post,* January 24, 1979. https://www.washingtonpost.com/archive/lifestyle/1979/01/24/doomsayer-of-the-me-decade/57f252a0-5a2b-4cdc-a9ff-c432e9bb2767/?noredirect=on&utm_term=.de772465d26e.

Allen, T. F. H., and Thomas Hoekstra. *Toward a Unified Ecology.* New York: Columbia University Press, 1992.

Ashby, W. Ross, *Design for a Brain: The Origin of Adaptive Behaviour.* 2nd rev. ed. New York: John Wiley & Sons, 1960.

Ashley, Robert, dir. *Music with Roots in the Aether: Pauline Oliveros.* New York: Lovely Music, 1987.

———. *Music with Roots in the Aether: Terry Riley.* New York: Lovely Music, 1987.

Backster, Cleve. *Primary Perception: Biocommunication with Plants, Living Foods, and Human Cells.* Anza, CA: White Rose Millennium Press, 2003.

Banes, Sally. *Terpsichore in Sneakers: Post-Modern Dance.* Boston: Houghton Mifflin, 1980.

Barbour, Ian, ed. *Western Man and Environmental Ethics: Attitudes toward Nature and Technology.* Reading, MA: Addison-Wesley, 1973.

Bataille, Gretchen M. *Native American Representations: First Encounters, Distorted Images, and Literary Appropriations.* Lincoln: University of Nebraska Press, 2001.

Bateson, Gregory. *Mind and Nature: A Necessary Unity.* New York: Dutton, 1979.

———. *Steps to an Ecology of Mind: Collected Essays in Anthropology, Psychiatry, Evolution, and Epistemology.* Chicago: University of Chicago Press, 1972.

Belgrad, Daniel. *The Culture of Spontaneity: Improvisation and the Arts in Postwar America.* Chicago: University of Chicago Press, 1998.

Berman, Morris. *The Reenchantment of the World.* Ithaca, NY: Cornell University Press, 1981.

Bernstein, David, ed. *The San Francisco Tape Music Center: 1960s Counterculture and the Avant-Garde.* Berkeley: University of California Press, 2008.

Bertalanffy, Ludwig von. *General System Theory: Foundations, Development, Applications.* Rev. ed. New York: George Braziller, 1968.

Boulding, Kenneth. "The Economics of the Coming Spaceship Earth." In *Environmental Quality in a Growing Economy: Essays from the Sixth RFF Forum,* ed. Henry Jarrett, 3–14. Baltimore: Johns Hopkins University Press, 1966.

Brand, Stewart. "Both Sides of the Necessary Paradox." *Harper's,* November 1973, 20–37.

Brannaman, Buck, and William Reynolds. *Faraway Horses: Adventures and Wisdom of an American Horse Whisperer.* New York: Lyons, 2003.

Breitweiser, Sabine, ed. *Simone Forti: Thinking with the Body.* Salzburg: Museum der Moderne, 2014.

Brinkema, Eugenie. *The Forms of the Affects.* Durham, NC: Duke University Press, 2014.

Brockman, John, ed. *The Third Culture: Beyond the Scientific Revolution.* New York: Simon & Schuster, 1995.

Brown, Barbara B., *New Mind, New Body: Bio-Feedback: New Directions for the Mind.* New York: Harper & Row, 1974.

Brown, Tom. *The Tracker: The Story of Tom Brown, Jr.* As told to William Jon Watkins. Englewood Cliffs, NJ: Prentice-Hall, 1978.

Burnett, D. Graham. *The Sounding of the Whale: Science and Cetaceans in the Twentieth Century.* Chicago: University of Chicago Press, 2012.

Cage, John. *Silence: Lectures and Writings.* London: Marion Boyars, 1995.

Caplan, Arthur, ed. *The Sociobiology Debate: Readings on Ethical and Scientific Issues.* New York: Harper & Row, 1978.

Capra, Fritjof. *The Turning Point: Science, Society, and the Rising Culture.* New York: Bantam, 1982.

Carson, Rachel. *Silent Spring.* New York: Fawcett World Library, 1962.

Clayton, Martin. *Time in Indian Music: Rhythm, Metre, and Form in North Indian Rāg Performance.* New York: Oxford University Press, 2008.

Clifton, James, ed. *The Invented Indian: Cultural Fictions and Government Policies.* New Brunswick, NJ: Transaction, 1990.

Collins, Nicolas. "'Pea Soup'—a History." Unpublished manuscript. http://www.nicolascollins.com/texts/peasouphistory.pdf.

Commoner, Barry. *Science and Survival.* New York: Viking Press, 1966.

———. *The Closing Circle: Nature, Man, and Technology.* New York: Knopf, 1971.
Cousteau, Jacques-Yves, and Philippe Diolé. *Dolphins*, trans. J. F. Bernard. Garden City, NY: Doubleday, 1975.
Cox, Christoph, and Daniel Warner, eds. *Audio Culture: Readings in Modern Music.* New York: Continuum, 2004.
Dawkins, Richard. *The Selfish Gene.* New York: Oxford University Press, 1989.
Daws, Gavan. "'Animal Liberation' as Crime: The Hawaii Dolphin Case." In *Ethics and Animals*, ed. Harlan Miller and William H. Williams. Clifton, NJ: Humana Press, 1983.
Deloria, Philip J. *Playing Indian.* New Haven, CT: Yale University Press, 1998.
Deloria, Vine, Jr. *God Is Red: A Native View of Religion.* New York: Dell, 1973.
———. *The Metaphysics of Modern Existence.* New York: Harper & Row, 1979.
Diamond, Stephen. *What the Trees Said: Life on a New Age Farm.* New York: Delacorte Press, 1971.
Disch, Robert, ed. *The Ecological Conscience: Values for Survival.* Englewood Cliffs, NJ: Prentice-Hall, 1970.
Duckworth, William. *Talking Music: Conversations with John Cage, Philip Glass, Laurie Anderson, and Five Generations of American Experimental Composers.* New York: Schirmer, 1995.
Edwards, Paul N. *The Closed World: Computers and the Politics of Discourse in Cold War America.* Cambridge, MA: MIT Press, 1996.
Egler, Frank. "Pesticides—In Our Ecosystem." *American Scientist* 52, no. 1 (March 1964): 110–36.
Ehrlich, Paul. *The Population Bomb.* New York: Ballantine, 1968.
Ellsberg, Daniel. *Papers on the War.* New York: Simon & Schuster, 1972.
Eno, Brian. *Imaginary Landscapes.* New York: Mystic Fire Video, 1989.
———. *A Year with Swollen Appendices: Brian Eno's Diary.* London: Faber and Faber, 1996.
Eno, Brian, and Russell Mills. *More Dark than Shark.* London: Faber and Faber, 1986.
Forti, Simone. "Dancing at the Fence." *Avalanche* 10 (December 1974): 20–23.
———. *Handbook in Motion.* Northampton, MA: Contact Editions, 1974.
Fuller, R. Buckminster. *Operating Manual for Spaceship Earth.* Carbondale: Southern Illinois University Press, 1969.
Gaskin, Stephen. *Haight Ashbury Flashbacks.* Berkeley: Ronin, 1990.
Ghamari-Tabrizi, Sharon. *The Worlds of Herman Kahn: The Intuitive Science of Thermonuclear War.* Cambridge, MA: Harvard University Press, 2005.
Goodall, Jane, *In the Shadow of Man.* Boston: Houghton Mifflin, 1971.
Gould, Stephen Jay. "Is a New and General Theory of Evolution Emerging?" *Paleobiology* 6, no. 1 (Winter 1980): 119–30.
Gray, Elizabeth Dodson. *Green Paradise Lost.* Wellesley, MA: Roundtable Press, 1979.
Greenpeace: Voyages to Save the Whales. Canada: Greenpeace Film Productions, 1977.

Griffin, Donald R. *The Question of Animal Awareness: Evolutionary Continuity of Mental Experience*. New York: Rockefeller University Press, 1976.

Grimshaw, Jeremy. *Draw a Straight Line and Follow It: The Music and Mysticism of La Monte Young*. New York: Oxford University Press, 2011.

Hagen, Joel B. *An Entangled Bank: The Origins of Ecosystem Ecology*. New Brunswick, NJ: Rutgers University Press, 1992.

Halper, Jon, ed. *Gary Snyder: Dimensions of a Life*. San Francisco: Sierra Club Books, 1991.

Halpern, Orit. *Beautiful Data: A History of Vision and Reason since 1945*. Durham, NC: Duke University Press, 2014.

Harvey, David. *A Brief History of Neoliberalism*. Cambridge: Oxford University Press, 2005.

Hayles, N. Katherine. *How We Became Posthuman: Virtual Bodies in Cybernetics, Literature, and Informatics*. Chicago: University of Chicago Press, 1999.

Heims, Steve J. *The Cybernetics Group*. Cambridge, MA: MIT Press, 1991.

———. *John von Neumann and Norbert Wiener: From Mathematics to the Technologies of Life and Death*. Cambridge, MA: MIT Press, 1980.

Hine, Thomas. *The Great Funk: Falling Apart and Coming Together (on a Shag Rug) in the Seventies*. New York: Farrar, Straus and Giroux, 2007.

Horowitz, Daniel, ed. *Jimmy Carter and the Energy Crisis of the 1970s: "The Crisis of Confidence" Speech of July 15, 1979*. New York: Bedford/St. Martin's, 2005.

Hunt, Ray. *Think Harmony with Horses: An In-Depth Study of Horse/Man Relationship*. Bruneau, ID: Give-It-A-Go Books, 1978.

Hunter, Robert. *Warriors of the Rainbow: A Chronicle of the Greenpeace Movement from 1971 to 1979*. Fremantle, WA: Greenpeace in association with Fremantle Press, 2011 (1979).

Jantsch, Erich. *The Self-Organizing Universe: Scientific and Human Implications of the Emerging Paradigm of Evolution*. New York: Pergamon Press, 1980.

Jenkins, Philip. *Decade of Nightmares: The End of the Sixties and the Making of Eighties America*. New York: Oxford University Press, 2006.

Jervis, John. "Uncanny Presences." In *Uncanny Modernity: Cultural Theories, Modern Anxieties*, ed. Jo Collins and John Jervis, 10–50. New York: Palgrave Macmillan, 2008.

Johnson, Roger, ed. *Scores: An Anthology of New Music*. New York: Schirmer, 1981.

Kahn, Herman. *On Thermonuclear War*. New York: Taylor and Francis, 2017 (1960).

Kaiser, David, and W. Patrick McCray, eds. *Groovy Science: Knowledge, Innovation, and American Counterculture*. Chicago: University of Chicago Press, 2016.

Killen, Andreas. *1973 Nervous Breakdown: Watergate, Warhol, and the Birth of Post-Sixties America*. New York: Bloomsbury, 2006.

Kingsland, Sharon E. *The Evolution of American Ecology, 1890–2000*. Baltimore: Johns Hopkins University Press, 2005.

———. *Modeling Nature: Episodes in the History of Population Ecology.* Chicago: University of Chicago Press, 1985.

Kirk, Andrew. *Counterculture Green: The Whole Earth Catalog and American Environmentalism.* Lawrence: University Press of Kansas, 2007.

Koestler, Arthur, and J. R. Smythies, eds. *Beyond Reductionism: New Perspectives in the Life Sciences.* Boston: Beacon Press, 1969.

Kostelanetz, Richard, ed. *John Cage.* New York: Praeger, 1970.

Kostelanetz, Richard, and Robert Flemming, eds. *Writings on Glass: Essays, Interviews, Criticism.* New York: Schirmer Books, 1997.

Kristeva, Julia. *The Kristeva Reader,* ed. Toril Moi. New York: Columbia University Press, 1986.

LaBelle, Brandon. *Background Noise: Perspectives on Sound Art.* New York: Continuum, 2006.

Lasch, Christopher. *The Culture of Narcissism: American Life in an Age of Diminishing Expectations.* New York: Norton, 1979.

———. "The Narcissist Society." *New York Review of Books,* September 30, 1976. https://www.nybooks.com/articles/1976/09/30/the-narcissist-society/.

Laszlo, Ervin. *Introduction to Systems Philosophy: Toward a New Paradigm of Contemporary Thought.* New York: Gordon and Breach, 1972.

———. *Simply Genius! And Other Tales from My Life: An Informal Autobiography.* Carlsbad, CA: Hay House, 2011.

———. *The Systems View of the World.* New York: George Braziller, 1972.

Lavezzoli, Peter. *The Dawn of Indian Music in the West.* New York: Continuum, 2006.

Leopold, Aldo. *A Sand County Almanac.* New York: Oxford University Press, 1966.

Lewontin, R. C., Steven Rose, and Leon J. Kamin. *Not in Our Genes: Biology, Ideology, and Human Nature.* New York: Pantheon, 1984.

Lilly, John. *The Center of the Cyclone: An Autobiography of Inner Space.* New York: Bantam, 1972.

———. *The Mind of the Dolphin: A Nonhuman Intelligence.* Garden City, NY: Doubleday, 1967.

Linden, Eugene. *Silent Partners: The Legacy of the Ape Language Experiments.* New York: Times Books, 1986.

Lipset, David. *Gregory Bateson: The Legacy of a Scientist.* Englewood Cliffs, NJ: Prentice-Hall, 1980.

Lovelock, James. *Gaia: A New Look at Life on Earth.* New York: Oxford University Press, 1979.

Lubow, Arthur. "Riot in Fish Tank 11." *New Times,* October 14, 1977, 36–53.

Lucier, Alvin. *Reflections: Interviews, Scores, Writings.* Köln: MusikTexte, 1995.

Lucier, Alvin, and Douglas Simon. *Chambers: Scores by Alvin Lucier.* Middletown, CT: Wesleyan University Press, 2012.

Marder, Michael. *Plant-Thinking: A Philosophy of Vegetal Life.* New York: Columbia University Press, 2013.

Marder, Michael, and Luce Irigaray. *Through Vegetal Being: Two Philosophical Perspectives*. New York: Columbia University Press, 2016.

Margalef, Ramón. *Perspectives in Ecological Theory*. Chicago: University of Chicago Press, 1968.

Maynard Smith, John. *Evolution and the Theory of Games*. New York: Cambridge University Press, 1982.

McCloskey, J. Michael. *In the Thick of It: My Life in the Sierra Club*. Washington, DC: Island Press, 2005.

McHarg, Ian. *Design with Nature*. Garden City, NY: Natural History Press, 1969.

McIntosh, Robert. *The Background of Ecology: Concept and Theory*. Cambridge: Cambridge University Press, 1985.

McIntyre, Joan. *Mind in the Waters: A Book to Celebrate the Consciousness of Whales and Dolphins*. San Francisco: Sierra Club Books, 1974.

McKenna, Dennis. *The Brotherhood of the Screaming Abyss: My Life with Terence McKenna*. St. Cloud, MN: North Star Press, 2012.

McKenna, Terence. *True Hallucinations: Being an Account of the Author's Extraordinary Adventures in the Devil's Paradise*. San Francisco: HarperSanFrancisco, 1993.

Mirowski, Philip. *Machine Dreams: Economics Becomes a Cyborg Science*. New York: Cambridge University Press, 2002.

Mitchell, John, ed. *Ecotactics: The Sierra Club Handbook for Environmental Activists*. New York: Pocket Books, 1970.

Mori, Masahiro. "The Uncanny Valley," trans. Karl F. MacDorman and Norri Kageki. IEEE Spectrum. https://spectrum.ieee.org/automaton/robotics/humanoids/the-uncanny-valley. Updated June 12, 2012.

Morlock, Forbes. "Doubly Uncanny: An Introduction to 'On the Psychology of the Uncanny.'" *Angelaki* 2, no. 1 (1997): 17–21.

Mowat, Farley. *The Desperate People*. Rev. ed. Toronto: McClelland & Stewart, 1975.

———. *Never Cry Wolf*. New York: Bantam, 1973.

———. *People of the Deer*. Rev. ed. New York: Carroll & Graf, 1975.

———. *A Whale for the Killing*. Boston: Little, Brown, 1972.

Næss, Arne. "The Shallow and the Deep, Long-Range Ecology Movement: A Summary." *Inquiry* 16, no. 1 (February 1973): 95–100. http://dx.doi.org/10.1080/00201747308601682.

Nash, Roderick. *The Rights of Nature: A History of Environmental Ethics*. Madison: University of Wisconsin Press, 1989.

———. *Wilderness and the American Mind*. 3rd ed. New Haven, CT: Yale University Press, 1982.

Neuhaus, Max. Liner notes for *Max Neuhaus, Fontana Mix—Feed (Six Realizations of John Cage)*. Milan, Italy: Alga Marghen, 2003.

Neuhaus, Max, and Christoph Cox, eds. *Max Neuhaus*. New York: Dia Center for the Arts, 2010.

Norris, Kenneth S. *The Porpoise Watcher*. New York: Norton, 1974.

Novack, Cynthia. *Sharing the Dance: Contact Improvisation and American Culture.* Madison: University of Wisconsin Press, 1990.

Nyman, Michael. *Experimental Music: Cage and Beyond.* 2nd ed. New York: Cambridge University Press, 1999.

Oates, David. *Earth Rising: Ecological Belief in an Age of Science.* Corvallis: Oregon State University Press, 1989.

O'Barry, Richard. *Behind the Dolphin Smile.* Chapel Hill, NC: Algonquin Books, 1988.

Odum, Howard T., and Richard C. Pinkerton. "Time's Speed Regulator: The Optimum Efficiency for Maximum Power Output in Physical and Biological Systems." *American Scientist* 43, no. 2 (April 1955): 331–43.

Oliveros, Pauline. *Deep Listening: A Composer's Sound Practice.* New York: Universe, 2005.

———. *Software for People: Collected Writings, 1963–1980.* Baltimore: Smith Publications, 1984.

———. *Sonic Meditations.* Baltimore: Smith Publications, 1974.

Payne, Roger. "Humpbacks: Their Mysterious Songs." *National Geographic,* January 1979, 18–25.

Pickering, Andrew. *The Cybernetic Brain: Sketches of Another Future.* Chicago: University of Chicago Press, 2010.

Potter, Keith. *Four Musical Minimalists: La Monte Young, Terry Riley, Steve Reich, Philip Glass.* New York: Cambridge University Press, 2000.

Poundstone, William. *Prisoner's Dilemma: John von Neumann, Game Theory, and the Puzzle of the Bomb.* New York: Doubleday, 1992.

Pryor, Karen. "Behavior and Learning in Porpoises and Whales." *Naturwissenschaften* 60 (1973): 412–20.

———. *Lads Before the Wind: Adventures in Porpoise Training.* New York: Harper & Row, 1975.

———. "Why Porpoise Trainers Are Not Dolphin Lovers: Real and False Communication in the Operant Setting." *Annals of the New York Academy of Sciences* 364, no. 1 (1981): 137–43.

Pryor, Karen, Richard Haag, and Joseph O'Reilly. "The Creative Porpoise: Training for Novel Behavior." *Journal of the Experimental Analysis of Behavior* 12, no. 4 (July 1969): 654–61.

Retallack, Dorothy. *The Sound of Music and Plants.* Santa Monica, CA: DeVorss, 1973.

Riley, Terry. "Obsessed and Passionate about All Music," interviewed by Frank Oteri. February 16, 2001. https://2104310a1da50059d9c5-d1823d6f516b5299e7df5375e9cf45d2.ssl.cf2.rackcdn.com/nmbx/assets/26/riley_interview.pdf.

Rogin, Michael. *"Ronald Reagan," the Movie: And Other Episodes in Political Demonology.* Berkeley: University of California Press, 1987.

Rome, Adam. *The Genius of Earth Day: How a 1970 Teach-In Unexpectedly Made the First Green Generation.* New York: Hill & Wang, 2013.

Ruckert, George E. *Music in North India: Experiencing Music, Expressing Culture.* New York: Oxford University Press, 2004.

Sabin, Paul. *The Bet: Paul Ehrlich, Julian Simon, and Our Gamble Over Earth's Future.* New Haven, CT: Yale University Press, 2013.

Sahlins, Marshall. *The Use and Abuse of Biology: An Anthropological Critique of Sociobiology.* Ann Arbor: University of Michigan Press, 1976.

Schafer, R. Murray. *The Book of Noise.* Vancouver: Price Print, 1970.

———. *Ear Cleaning: Notes for an Experimental Music Course.* Toronto: Berandol Music Limited, 1967.

———. *The New Soundscape: A Handbook for the Modern Music Teacher.* Don Mills, Ontario: BMI Canada, 1969.

———. *The Tuning of the World.* New York: Knopf, 1977.

Schulman, Bruce J. *The Seventies: The Great Shift in American Culture, Society, and Politics.* New York: Free Press, 2001.

Sears, Paul B. *Deserts on the March.* Norman: University of Oklahoma Press, 1935.

———. "Ecology—A Subversive Subject." *BioScience* 14, no. 7 (1964): 11–13.

The Secret Life of Plants, dir. Walon Green. Infinite Enterprises, 1978.

Sedgwick, Eve Kosofsky. *Touching Feeling: Affect, Pedagogy, Performativity.* Durham, NC: Duke University Press, 2003.

Segerstråle, Ullica. *Defenders of the Truth: The Battle for Science in the Sociobiology Debate and Beyond.* New York: Oxford University Press, 2000.

Shepard, Paul. *Thinking Animals: Animals and the Development of Human Intelligence.* New York: Viking, 1978.

Shepard, Paul, and Daniel McKinley, eds. *The Subversive Science: Essays towards an Ecology of Man.* Boston: Houghton Mifflin, 1970.

Sheppard, David. *On Some Faraway Beach: The Life and Times of Brian Eno.* Chicago: Chicago Review Press, 2009.

Simon, Julian. *The Ultimate Resource.* Princeton, NJ: Princeton University Press, 1981.

Simon, Julian, and Herman Kahn, eds. *The Resourceful Earth: A Response to Global 2000.* New York: Blackwell, 1984.

Singer, Peter. *Animal Liberation: A New Ethics for Our Treatment of Animals.* New York: New York Review of Books, 1975.

Snyder, Gary. *The Gary Snyder Reader.* Washington, DC: Counterpoint, 1999.

———. *The Old Ways: Six Essays.* San Francisco: City Lights Books, 1977.

———. *The Real Work: Interviews and Talks, 1964–1979.* New York: New Directions, 1980.

———. *Turtle Island.* New York: New Directions, 1974.

Stone, Christopher D. "Should Trees Have Standing? Toward Legal Rights for Natural Objects." *Southern California Law Review* 45 (1972): 450–501.

Tamm, Eric. *Brian Eno: His Music and the Vertical Color of Sound.* Boston: Faber and Faber, 1989.

Tart, Charles T., ed. *Altered States of Consciousness*. 2nd ed. New York: Anchor Books, 1972.

Tobias, Michael, ed. *Deep Ecology*. San Diego: Avant Books, 1988.

Tompkins, Peter, and Christopher Bird. *The Secret Life of Plants*. New York: Harper, 1973.

Tucker, William, *Progress and Privilege: America in the Age of Environmentalism*. Garden City, NY: Anchor, 1982.

Turner, Fred. *The Democratic Surround: Multimedia and American Liberalism from World War II to the Psychedelic Sixties*. Chicago: University of Chicago Press, 2013.

———. *From Counterculture to Cyberculture: Stewart Brand, the Whole Earth Network, and the Rise of Digital Utopianism*. Chicago: University of Chicago Press, 2006.

Udall, Stewart. *The Quiet Crisis*. New York: Holt, Rinehart, and Winston, 1963.

Vizenor, Gerald Robert. *Fugitive Poses: Native American Indian Scenes of Absence and Presence*. Lincoln: University of Nebraska Press, 2000.

Weil, Andrew. *The Marriage of the Sun and Moon: A Quest for Unity in Consciousness*. Boston: Houghton Mifflin, 1980.

———. *The Natural Mind: A New Way of Looking at Drugs and the Higher Consciousness*. Boston: Houghton Mifflin, 1972.

Weil, Andrew, and Peter Fremont. "Interview with Andrew Weil." *High Times* 1, no. 5 (August–September 1975): 15–17, 58.

Weiss, Paul A. *The Science of Life: The Living System—A System for Living*. Mt. Kisco, NY: Futura, 1973.

———. "The System of Nature and the Nature of Systems." In *Toward a Man-Centered Medical Science*, ed. Karl E. Schaefer, Herbert Hensel, and Ronald Brady, 17–63. Mt. Kisco, NY: Futura, 1977.

Wells, Tom. *Wild Man: The Life and Times of Daniel Ellsberg*. New York: Palgrave, 2001.

Weyler, Rex. *Greenpeace: How a Group of Journalists, Ecologists, and Visionaries Changed the World*. Emmaus, PA: Rodale, 2004.

Wiener, Norbert. *Cybernetics: Or Control and Communication in the Animal or Machine*. 2nd ed. Cambridge, MA: MIT Press, 1961.

———. *God and Golem, Inc.: A Comment on Certain Points Where Cybernetics Impinges on Religion*. Cambridge, MA: MIT Press, 1964.

———. *The Human Use of Human Beings: Cybernetics and Society*. 1st ed. Boston: Houghton Mifflin, 1950.

———. *The Human Use of Human Beings: Cybernetics and Society*. 2nd rev. ed. Boston: Houghton Mifflin, 1954.

Wilden, Anthony. *System and Structure: Essays in Communication and Exchange*. 2nd ed. New York: Tavistock, 1980.

Willers, Bill, ed. *Learning to Listen to the Land*. Washington, DC: Island Press, 1991.

Wilson, Edward O. *On Human Nature*. Cambridge, MA: Harvard University Press, 1978.

———. *Sociobiology: The New Synthesis*. Cambridge, MA: Harvard University Press, 1975.

Wolfe, Tom. "The 'Me' Decade and the Third Great Awakening." *New York Magazine*, August 23, 1976. http://nymag.com/news/features/45938/.

Zimmermann, Walter. *Desert Plants: Conversations with 23 American Musicians*. Vancouver: A.R.C. Publications, 1976.

Index